OCT 0 3 2016

Writing for Inventors

D0754207

Fritzing for Inventors

Take Your Electronics Project from Prototype to Product

Simon Monk

New York   Chicago   San Francisco
Athens   London   Madrid
Mexico City   Milan   New Delhi
Singapore   Sydney   Toronto

McGraw-Hill Education books are available at special quantity discounts to use as premiums and sales promotions or for use in corporate training programs. To contact a representative, please visit the Contact Us page at www.mhprofessional.com.

Fritzing for Inventors: Take Your Electronics Project from Prototype to Product

Copyright © 2016 by McGraw-Hill Education. All rights reserved. Printed in the United States of America. Except as permitted under the United States Copyright Act of 1976, no part of this publication may be reproduced or distributed in any form or by any means, or stored in a database or retrieval system, without the prior written permission of the publisher.

1 2 3 4 5 6 7 8 9 0   DOC/DOC   1 2 0 9 8 7 6 5

ISBN 978-0-07-184463-5
MHID   0-07-184463-5

This book is printed on acid-free paper.

Sponsoring Editor Michael McCabe	**Copy Editor** Patti Scott
Editing Supervisor Donna M. Martone	**Proofreader** Claire Splan
Production Supervisor Pamela A. Pelton	**Indexer** Claire Splan
Acquisitions Coordinator Amy Stonebraker	**Art Director, Cover** Jeff Weeks
Project Manager Patricia Wallenburg, TypeWriting	**Composition** TypeWriting

McGraw-Hill Education, the McGraw-Hill Education logo, TAB, and related trade dress are trademarks or registered trademarks of McGraw-Hill Education and/or its affiliates in the United States and other countries and may not be used without written permission. All other trademarks are the property of their respective owners. McGraw-Hill Education is not associated with any product or vendor mentioned in this book.

Information contained in this work has been obtained by McGraw-Hill Education from sources believed to be reliable. However, neither McGraw-Hill Education nor its authors guarantee the accuracy or completeness of any information published herein, and neither McGraw-Hill Education nor its authors shall be responsible for any errors, omissions, or damages arising out of use of this information. This work is published with the understanding that McGraw-Hill Education and its authors are supplying information but are not attempting to render engineering or other professional services. If such services are required, the assistance of an appropriate professional should be sought.

To Matthew,
from a very proud Dad.

About the Author

Dr. Simon Monk (Preston, UK) has a degree in cybernetics and computer science and a Ph.D. in software engineering. He spent several years as an academic before he returned to industry, co-founding the mobile software company, Momote Ltd. He has been an active electronics hobbyist since his early teens and is a full-time writer on hobby electronics and open-source hardware. Dr. Monk is the author of numerous electronics books, specializing in open-source hardware platforms, especially Arduino and Raspberry Pi. He is also co-author with Paul Scherz of *Practical Electronics for Inventors, Third Edition.*

You can follow him on Twitter, where he is @simonmonk2.

Contents

Acknowledgments

Many thanks to all those at McGraw-Hill who have done such a great job in producing this book. In particular, thanks to my editors Roger Stewart and Michael McCabe, and to Patty Wallenburg.

I am most grateful to André Knörig for his technical review of the material.

And, last but not least, thanks once again to Linda, for her patience and generosity in giving me space to do this.

CHAPTER 1

Introduction to Fritzing

For many makers and electronics hobbyists, the idea of using computer-aided design (CAD) software conjures up images of obscure and difficult-to-use programs that might need a week's training course before you can do anything useful with them. Fritzing is not like that. Fritzing has been designed and has developed for makers, hobbyists, and inventors, not for professional electronic engineers.

This high-speed tour of Fritzing will give you an idea of the breadth of this tool. It is useful just to know what features are available, even if you are not going to use them right away. Many of the things mentioned in the following short summary will be dealt with in considerably greater detail later in the book, so please don't feel cheated.

Fritzing

Fritzing is so easy to use that many people use it to sketch out breadboard layouts or draw schematic diagrams, as this can be accomplished almost as easily as with pen and paper. This ease of use does not mean that Fritzing is only of use for simple example projects. It is perfectly possible to design quite complex projects with Fritzing, without having to compromise on the design. The key advantages of Fritzing over other CAD tools are as follows:

- It is free to use (although a donation is appreciated).
- It is simple and intuitive.
- Fritzing includes many libraries of components from popular suppliers such as Adafruit, Sparkfun, and Snootlabs.

- It is suitable for Arduino, Raspberry Pi, Beaglebone, and Spark Core projects.
- Fritzing is an integrated printed circuit board (PCB) production service. Or export the files and use another service.

Figure 1-1 shows one of the example projects supplied with Fritzing.

The first thing that may come as a surprise, if you have used other CAD systems, is that the first view of the project is a Breadboard view. Fritzing assumes that you will want to make a prototype of your project on solderless breadboard to get all the wrinkles out of it before you move on to making a PCB. Most CAD systems completely ignore the concept of prototyping on breadboard, but in Fritzing, it is a fundamental part of the design process.

For many people, it can be simpler to visualize the project from the point of view of actual components and wires rather than from the more abstract schematic diagram. You don't have to start your Fritzing design with a breadboard layout. If you prefer, start with the schematic, but at least the option is there.

Although this example uses solderless breadboard, there are other prototyping options that you can use. Many people like to use stripboard or other types of

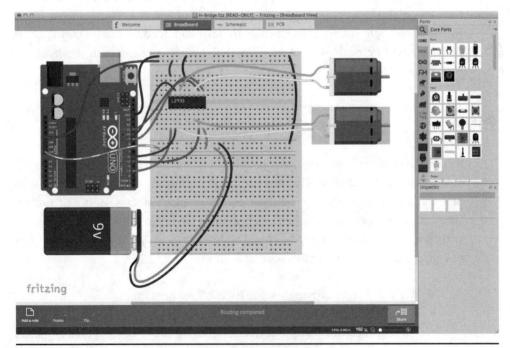

FIGURE 1-1   A Fritzing example project (Breadboard view).

premade prototyping PCB to construct soldered prototypes that are a little more robust than the solderless breadboard. If you want to take this approach, then from within the Breadboard view, you just delete the breadboard and add the type of prototyping board you want. Other sizes and styles of breadboard are also available in the parts library.

If you click on the Schematic tab in this project, you will see the schematic diagram in Figure 1-2.

This represents the same design as the Breadboard view, but as a schematic diagram. If we were to add a new component here or delete one or connect the components differently, those changes would also be applied to the Breadboard tab, were we to switch over to that view.

In other words, the Breadboard and Schematic views are both views of the same underlying design. Each of these views has its own information, such as where the components appear on the screen, but fundamentally they are views of the same design that will always stay consistent with each other.

If we were to switch to the PCB tab, we would see that this design is actually for a plug-in shield for an Arduino microcontroller board, designed to control a pair of motors (Figure 1-3).

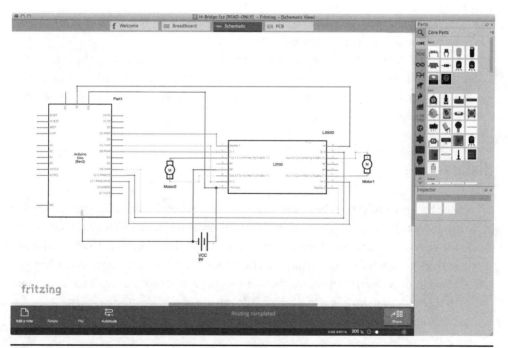

FIGURE 1-2 A Fritzing example project (Schematic view).

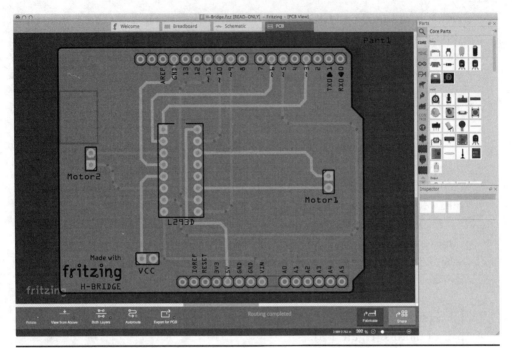

FIGURE 1-3 A Fritzing example project (PCB view).

The PCB view is linked to the underlying design in just the same way as the Breadboard and Schematic views.

In all three views, there are two areas on the right-hand side of the screen. The top area lists a lot of components and modules that you can add to your design. Below that is an information area that tells you about the currently selected component.

History

The Fritzing project has been developed by researchers at the University of Applied Sciences at Potsdam. The software is open-source and cross-platform. It is available for Windows, Mac, and Linux. This is a substantial project that has helped many makers get started designing and documenting their electronic projects. When you come to download the project, there is an option to donate a small sum of money to this very worthwhile project and help ensure its future development.

You can find out more about Fritzing at the project's website: http://fritzing.org.

Installing Fritzing

To install Fritzing, you will need a computer running Windows, Mac, or Linux. Windows users will need Windows XP or newer; Mac users, 10.7 and up; Linux users, any fairly recent distribution.

First download Fritzing from http://fritzing.org/download. Don't forget to donate! You will then be offered a download in a choice of formats, as shown in Figure 1-4.

Identifying your operating system is easy enough, but then you have to choose between 32-bit and 64-bit. These days, unless you have a computer that is more than three years old, you will almost certainly want the 64-bit version. If you have a Mac, you are very unlikely to have one old enough to require 32-bit while still running OSX 10.7 or later, so go for the 64-bit version.

If you pick the wrong version, you can always just download the other version. Fritzing does not use an installer program; it just downloads a zip file containing an executable program, which you can place wherever you like on your computer's file system. Typically, that will mean moving it to "Program files" on Windows or "Applications" on Mac.

Whatever your platform, all you have to do is double-click on the application program file to run it.

Version 0.9.0b was released on **July 14, 2014**.

Windows 32 bit

Windows 64 bit

Mac OS X 10.6 and up

Linux 32 bit

Linux 64 bit

Source

FIGURE 1-4 Fritzing download options.

Examples

The Fritzing installation includes a lot of example designs that you can open in Fritzing and either learn from them or save copies of them (use Save As...) to modify to make your own designs. These are accessible from the File | Open Example menu item. Some of the more interesting examples to peruse are as follows:

- H-bridge, the example used in Figures 1-1 to 1-3
- LED-Matrix, another Arduino design using a 5 × 7 LED matrix
- VoltageRegulator_7800series, which is a simple voltage regulator project

Documents and Designs

Editing a design in Fritzing is rather like working with a Word document. Everything relating to your design is held in a single file with the extension .fzz. Once Fritzing is installed, you can simply open a design by double-clicking on it. You can also keep the design anywhere that is convenient on your computer's filing system. Note that Fritzing uses the term *sketch* to mean a design.

Just as with most document-based systems, you can also use the File menu of Fritzing to Open, Save, or Save As... designs (sketches). You can also export various aspects of the design from the Export submenu on the File menu. This includes

- Saving the different design drawings to different image formats
- Generating industry standard files for PCB production
- Generating a list of parts used in the design, called a Bill of Materials, usually abbreviated BOM
- Exporting files to allow your design to be simulated in a circuit simulator that will read files in SPICE format

Going All the Way

Although this book will take you right from the initial schematic or breadboard prototype to the soldered PCB, there is still considerable benefit to using Fritzing without going any further than the breadboard layout. Although you can move things around on a real breadboard, you can't drag and drop a big selection of components from one area of the board to another in the way you can with the virtual board in Fritzing. Fritzing is also a great way of documenting your design and producing nice-looking pictures of your breadboard layout if you are going to write up and share your project on your blog or as an Instructable.

If you want to build a permanent one-off project using stripboard or some other kind of prototyping board, then Fritzing is an easy way to plan out where things will go on the prototyping board.

Using Views

The default appearance of the editor windows is pretty clean and easy on the eyes. However, if you wish, you can change background colors, turn the grid visibility on and off, and do various other things to get the views exactly as you like them.

When it comes to controlling what you can see in the window, you will need to use the zoom commands, either from the menu (View | Fit In Window is very useful) or from the slider on the bottom right of the editor area of the window.

The canvas where you add parts will automatically expand in any direction. So do not worry about leaving room for things; you can add them above or to the left of the existing extent of the diagram, and the size of the canvas will automatically expand to make room for them.

As far as moving things around, you can just click and drag parts, as you would expect. You can also click and drag or shift-click to select multiple items.

The Programming Window

One really interesting feature of Fritzing, that at the time of this writing is not quite ready yet, allows you to associate programs with a project. For example, if your design is creating an add-on board for an Arduino, such as the example we saw earlier, you might want to have an Arduino test program that will demonstrate how the shield should be used.

This feature would be even more useful if you were creating a project based on an Arduino for which the Arduino code was an essential part. By keeping it with the Fritzing design, you avoid many of the version management pitfalls where you might end up installing mismatched software on your hardware design.

Figure 1-5 shows this programming window in version 0.8.7 of Fritzing.

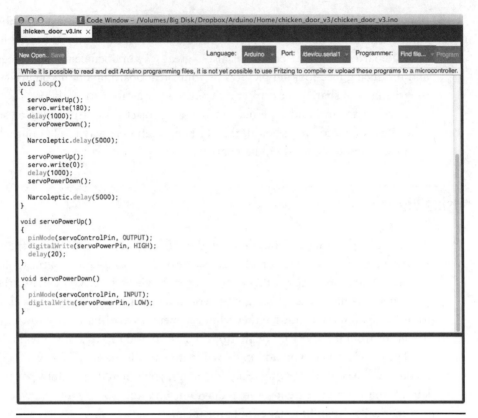

Figure 1-5   The Fritzing programming window.

Currently, this version allows you to edit code for Arduino and PicAxe, but the feature to upload onto the device is currently available for only PicAxe.

Autorouting

When you are in the PCB view, the Routing menu will offer up all sorts of options relating to the layout of a PCB. Fritzing includes an autorouter. This will attempt to route the copper tracks on the PCB from the leg of one component to another without the copper tracks crossing or getting too close to one another.

Generally, in a design you might use a combination of autorouting with a bit of manual tweaking to get the design just as you want it. Also, it can be quite fun trying to route the tracks by hand.

Autorouting and manual routing have a chapter of this book (Chapter 5) all to themselves.

Summary

I hope this chapter has whetted your appetite to get started with some Fritzing! If so, then the next chapter will appeal to you, as it is a quick-start chapter leading you through a whole miniproject from end to end.

CHAPTER 2

Quick Start

The best way to learn things is to try them yourself. So in this chapter you will be building an example project one step at a time. To get the most from this chapter, start Fritzing and follow the steps involved in making this simple light-emitting diode (LED) flasher project.

The project that we are going to design uses a 555 timer chip to flash two LEDs. The design is an extremely common one. If you search Google images for "555 timer LED flasher," you will find no end of similar designs.

The goal of this chapter is to get you started quickly and see how you can use Fritzing to take you all the way from a schematic to a PCB. As such, little attention is paid here to producing a nice-looking design or to doing things the "optimal" way. That's covered in later chapters as you go through each stage of the process in considerably greater detail.

You should also not worry too much about how this circuit works. In this chapter we are just interested in using Fritzing, not in electronic theory.

All the Fritzing files used in this book are available for download from the book's webpage at www.simonmonk.org. So if you wish, you can download the file and open it in Fritzing. The file for this chapter is called ch_02_Flasher.fzz.

Starting a New Project

When you first start Fritzing, a new project will automatically be created, ready for you to use. It will not be saved and it will have the name "Untitled Sketch.fzz." Start by saving the project somewhere handy, even though you have not done anything yet. A good name for the project would be "Flasher." Save it in a directory, perhaps in your Documents folder, so that you can find it again.

Once you have saved the file, you will notice that the name in the title bar has changed to "Flasher.fzz."

Drawing the Schematic

Some projects will be so simple and, for example, will use modular components such as an Arduino. For such projects, you may want to start with the Breadboard view. It makes sense to just think of modules and wires and breadboard rather than the more abstract Schematic view of the project. However, in this chapter you will start with the schematic view, as the project we are creating would be difficult to understand if we went straight to the Breadboard view. It also gives me an excuse to show you all the Editor views in action.

Finding a 555 Timer Part

Click on the Schematic tab. Not unsurprisingly, you will be greeted with a blank canvas. This is where you are going to add the components used in the project as component symbols and then connect them all.

When the schematic is finished, it will look like Figure 2-1. Actually, there is no power supply on the schematic as it stands. We will fix that in a later section.

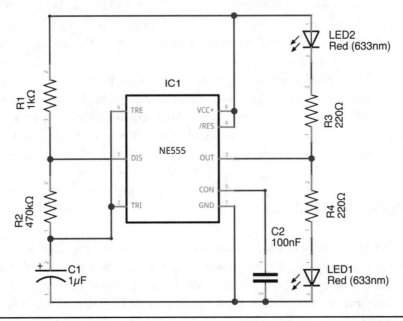

FIGURE 2-1   Final schematic.

From Figure 2-1, you can see that everything is centered on the 555 timer integrated circuit (IC). It makes sense to first add this component and then position the other components around it.

Fritzing has a lot of parts in its parts library and so the first problem is finding the IC we want in the parts library. You could try to browse for it, but it is often quicker just to use the search facility, so click in the "magnifying glass" icon at the top right and type 555 and press ENTER. The search results will then be shown in the Parts area (see Figure 2-2).

The first thing to notice is that you will have found more than one possible part. This is partly because some of the results are for other parts that happen to contain the number 555. There is, for example, a diode that has sneaked into the results. Some of the results are also breakout boards for using the surface-mount versions of the 555 chip on breadboard. You don't want any of these. You just need a regular 8-pin dual in-line (DIL) 555 timer chip.

You can find out more about each of the search results by hovering over it to refresh the Inspector area below the Parts area. From this it seems that three of the search results are what we are looking for. So why are there duplicates and which one should we use?

The answer to the first question is that different contributors to the library have added their own versions of the 555 timer part. To decide which one to use, drag the three candidates out onto the canvas and zoom in on them so that you can see them properly. You can later delete the parts that you don't want. Figure 2-3 shows the three alternatives.

The leftmost two parts appear to be identical, and if you look closely at them, you can see that the pins are identified by their pin number, whereas the right-hand part has meaningful names for the pins such as TRI (trigger), DIS (discharge),

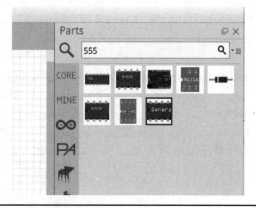

FIGURE 2-2   Searching for a 555 timer.

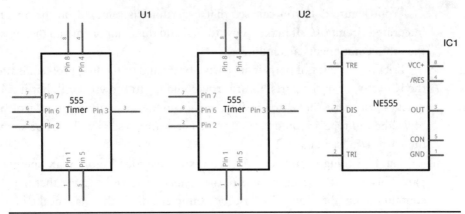

FIGURE 2-3   Alternative 555 timer parts.

etc. This is the most useful version of the 555 part, so use that one, deleting the other two by selecting them and pressing the BACKSPACE key.

Adding the Resistors

Resistors are such a common component that they should be easy to find, and indeed they are. Click on the CORE button in the Parts area of the screen, and there is a resistor, the first component in the list. Drag a resistor off the Parts area and onto the canvas to the left of the 555 timer (Figure 2-4).

The resistor has a default value of 220Ω. This particular resistor needs to be 1kΩ. So select the resistor so that the Inspector area underneath the parts list shows its properties. Change the value in the resistance to 1k either by typing **1k** or picking "1kΩ" from the drop-down list. Note that the label next to the resistor on the schematic has immediately updated to 1kΩ.

It would be better for the diagram if the resistor was vertical rather than horizontal, so with the resistor still selected, click on the Rotate button at the bottom of the window.

To add the three more resistors required by the design, you can "duplicate" the existing resistor by right-clicking on it and selecting "Duplicate" from the menu. This can be quicker than simply adding three resistors from the Parts area, because the resistor will already be the right way around. When you have added the three resistors, modify their values so that they agree with Figure 2-5.

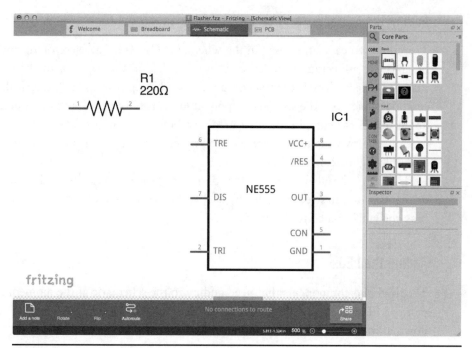

FIGURE 2-4   Adding a resistor.

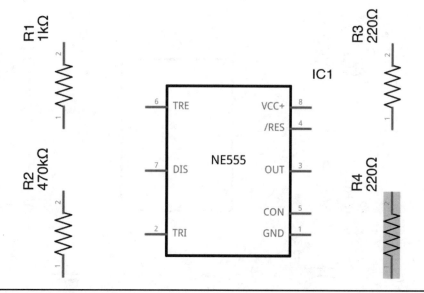

FIGURE 2-5   All the resistors added to the schematic.

Adding the Capacitors

There are two capacitors used in the schematic. One is a 1μF electrolytic and the other is a 10nF ceramic capacitor. These are also on the top row of the CORE parts. The electrolytic is the last component on the top row and is cylindrical. The ceramic capacitor is the second component in the list. Remember, to find out more about one of the parts in the Parts area, just hover your mouse over it.

Add the two capacitors, the electrolytic first, and the schematic should now resemble Figure 2-6.

If some of the labels overlap other parts, or are too far away from the part they belong to, you can select the label by itself (after selecting the part as a whole) and drag the label to a better location.

Adding the LEDs

The final two components that need adding to the schematic at the moment are a pair of LEDs. These are also under CORE components, but you will need to scroll down a little to find them. Position them as shown in Figure 2-7.

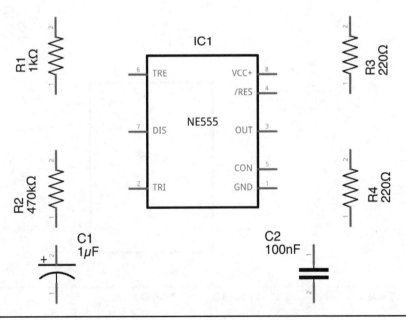

FIGURE 2-6   The capacitors added to the schematic.

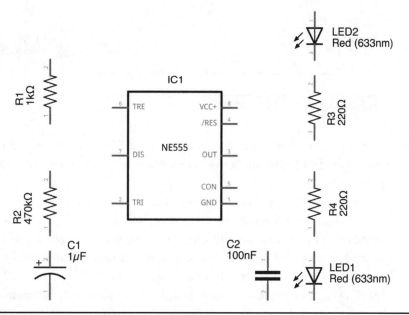

FIGURE 2-7   The LEDs added to the schematic.

Connecting Things

Now that you have all the parts on the canvas, you need to connect them. This can be a little trickier than you might expect the first time you try it.

Start by connecting the bottom of C1 to the bottom of C2. When you click on the bottom lead of C1, you will notice it will change color from dark red to another brighter color. This indicates that this is a good place to drag out a connection from one component lead to another. So drag out from the bottom connection of C1 (the dark red bit) toward the bottom lead of C2. When the two meet, the bottom lead of C2 will be highlighted to indicate that it's a good place to drop the end of the connection (Figure 2-8).

FIGURE 2-8   Connecting C1 and C2.

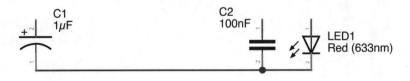

FIGURE 2-9   Connecting LED1 to C2.

Now do the same thing with LED1 and C2. This time drag from the bottom connection of LED1 to the bottom connection of C2. The result should look like Figure 2-9.

Notice how there is now a blob over the connection at the bottom of C2. This shows that all three connections are coming together at one place. It is important to check that these blobs appear when you have made a connection like this. For example, if we had dragged a connection in the opposite direction from C2 to LED1, it is easy to accidentally extend the connection out from C1 to LED1, bypassing C2 while making it look as if they are connected. But since there is no blob, you can tell that the connection is not as you would want. Figure 2-10 shows this faulty connection.

Next, connect the CON pin of the 555 timer to the top of C2. This will create a diagonal line from the pin of the chip to the capacitor, as shown in Figure 2-11a. While technically fine, schematic diagrams tend to have everything at right angles. So let's add a "bend point" to the line, by clicking anywhere along its length. A dot will appear on the line (Figure 2-11b). Drag this dot up to position the lines at right angles, as shown in Figure 2-11c.

Other connections can be tricky to make when you need to connect from the pin of a device to somewhere on a line. For example, you need to connect the OUT pin of the 555 timer to the line that connects R3 and R4. Figure 2-12 shows the steps needed to do this.

Start by connecting R3 to R4, as shown in Figure 2-12a. If you tried to drag out a connection from the OUT pin of the 555 timer to the new line, it would not connect properly (no blob). However, if you click on the middle of the newly created line and make a bend point (Figure 2-12b), then you can connect to this bend point from the OUT pin, as shown in Figure 2-12c.

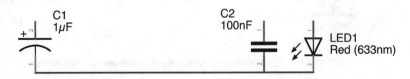

FIGURE 2-10   A bad connection (no blob).

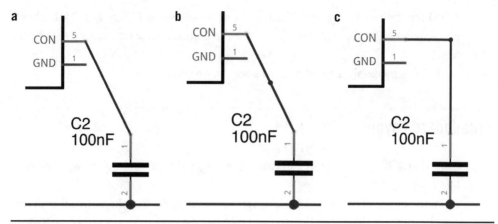

Figure 2-11   Adding a bend point to a line.

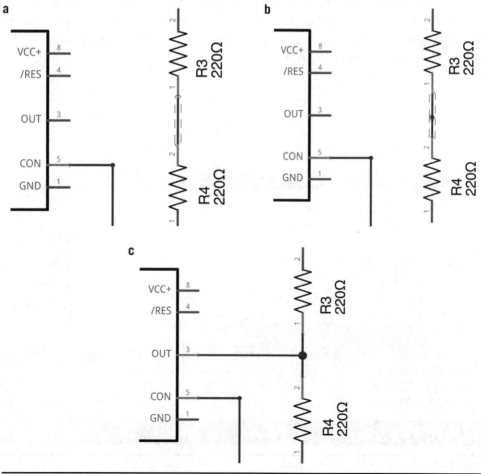

Figure 2-12   Connecting to an existing line.

Make the remainder of the connections as shown in Figure 2-1. Note that if you make a mistake, you can always use the Undo feature (CTRL-Z or ALT-Z on a Mac).

Now is probably a good time to save the project!

Breadboard Layout

Click on the Breadboard tab and you will be greeted by an unholy mess, something like Figure 2-13.

This isn't really a surprise. Although we have made the schematic all neat and tidy by putting the parts in the right places, we cannot expect this to happen automatically with the breadboard.

Notice that the components are connected with dashed lines. These are termed *rat's nest* in Fritzing, and after we have moved some of our components about, we can start turning these rat's nest connections into connections made on the breadboard or using jumper wires.

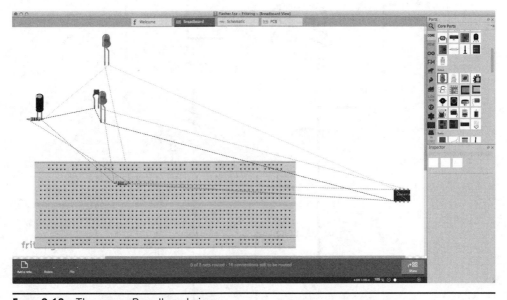

FIGURE 2-13 The messy Breadboard view.

Solderless Breadboard

Although Chapter 3 is devoted entirely to using solderless breadboard, if you have not used breadboard before, you will need to know a little more about how they work.

Figure 2-14 shows a small breadboard with a single resistor on it.

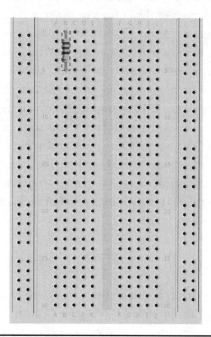

FIGURE 2-14   A half-size breadboard.

The breadboard has holes on the front, arranged in rows of five holes. Behind the plastic of the breadboard are clips that connect all those five holes. When you push a component's leads through the hole, it connects with the clip behind the hole. So in the example of Figure 2-14, the resistor is between rows 1 and 5 of the breadboard on the left-hand bank of rows. There is a similar bank of rows on the right-hand side of the breadboard. The two banks are not connected.

On either side of the breadboard are pairs of holes running vertically. These are marked with red and blue lines and are usually used as power connections. So you might connect GND to the blue column and the positive power supply to the red column. That way, when you need to connect one of the rows to GND or +V, you can use a jumper wire horizontally from the row you need to supply power to out to the nearest power bus column. You will see an example of this later.

Change the Breadboard

This is a pretty small design, so the full-size breadboard is too big. It would also be better in portrait than in landscape. So select the breadboard by clicking on it. Avoid clicking on it near a hole, or Fritzing will think you might be making a wire from that hole.

In the "size" field of the breadboard's properties, select the option "half+," then click the Rotate button at the bottom of the screen three times to rotate it through 270°. Click three times rather than once, because if you look closely at the breadboard you will see that it numbers the columns and we need to go through 270° for those numbers to start at the top of the board. Your breadboard layout is still a mess, but at least now we have the right breadboard for our project.

In looking at Figure 2-13, it appears that you have only two resistors rather than the four you should have. In fact, this is so because some of the resistors are stacked on top of one another. Gently drag them apart.

You will find this easiest to do if you drag the breadboard out of the way for now. When selecting a resistor that you think is on top of another, sometimes you accidentally select the rat's nest wire instead of the part. If this happens, just use the UNDO feature and try again.

Placing the Parts

The parts are now ready to put on the breadboard, starting with the 555 timer chip. Start by rotating this through 270°, as you did with the breadboard, so that pin 1 of the IC is at the top. There is a little dot next to pin 1 on the IC package. Drag the IC into the middle of the breadboard so that it straddles the two banks of rows.

As it meets the breadboard, you will see the rows connected to the pins become highlighted in green (Figure 2-15). By default, the parts do not show their names. To make it easier to identify them, right-click on each part and select the option "Show part label."

Before you add the parts to the breadboard, connect the GND pin of IC1 (pin 1) to the left-hand negative column on the breadboard and the VCC (+V) connection on pin 8 of IC1 (opposite pin 1) to the positive column. We will need some wires to do this, so drag out from the leftmost hole in the row that pin 1 of IC1 is connected to and drop the end of the wire across onto one of the holes on the negative column. Do the same thing for pin 8, but add a bend point in the wire to take it around the edge of IC1. Fritzing will pick a color for you. But generally it is the convention to make positive connections red and negative wires blue or black. So select each wire in turn and change its color, using the "color" drop-down list in its properties.

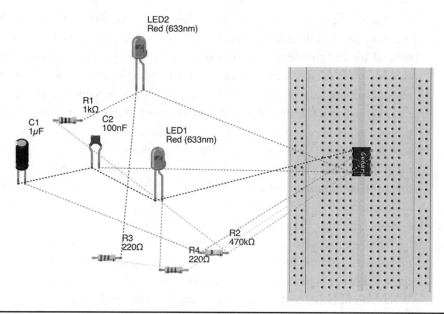

FIGURE 2-15 The IC in place on the breadboard.

We need most of the parts to be oriented vertically so that they can connect between rows on the breadboard. So rotate all the remaining components through 90°. Your design should now look something like Figure 2-16.

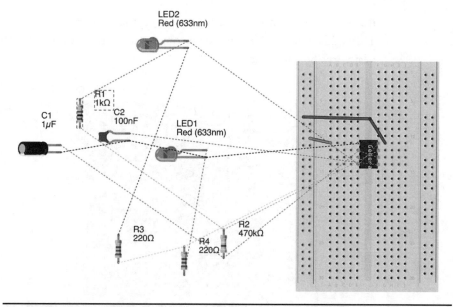

FIGURE 2-16 Parts rotated and power lines added.

Placing the remainder of the components is a bit like solving a puzzle. You want the parts to fit neatly on the breadboard with a spacing of the leads that is natural for the component. It is best to pick a component that you can see an obvious place for and add it. It can also help a lot to have the schematic printed out so that you can get another view on what should be connected to what.

Starting with C2, one of its connections needs to go to pin 5 of IC1, which is the bottom connection on the right-hand side. However, with C2 oriented as it stands, the capacitor body will be in the way, so select it, click the Flip button at the bottom of the window, and then place the capacitor so that its top pin is on the same breadboard row as pin 5 of the IC. Notice that when you drop C2 into the right place on the breadboard, the rat's nest wire from that pin of C2 has vanished, because that pin of C2 is now connected as it should be.

Connect the bottom pin of C2 to ground, using a new wire out to the negative breadboard column. Notice that as you start to drag the wire out from a hole, all the valid places where you could land the other end of the wire will turn yellow. When you have made the wire link, change its color to blue.

There is an obvious rat's nest line between pins 2 and 6 of the IC. Make this link with a wire and set the wire color to, say, orange. Figure 2-17 shows a close-up of the breadboard so far.

Now add the components in the following order. Remember to flip and rotate as you need to, in order to position the parts better. Use Figure 2-18 as a reference for placing the parts.

- **R2:** Drop the resistor so the top end is on the same row as pin 7 of IC1. Then add a wire back from the bottom pin up to pin 6 of IC1. Let the rat's nest wires be your guide.

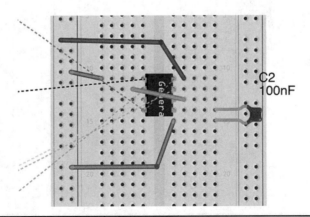

Figure 2-17 Breadboard with wires and C2.

- **R1:** Drop this so that its bottom end is on the same row as the top of R2 and pin 7 of the IC. Then drag a wire from the top pin of the resistor down to the same row as pin 8 of IC1.
- **C1:** Flip and then rotate twice before putting it on the breadboard with its pins connected to pins 1 and 2 of the IC.
- **R3:** The end of R3 that connects to pin 3 (OUT) of IC1 should go on the breadboard row for pin 3 of IC1. The other end of the resistor should then be four holes higher up the board.
- **R4:** This is the current-limiting resistor for the other LED, so place this next to R3 but heading down the board.
- **LED1:** This LED has one pin to GND and the other to the free end of R4, so rotate it, flip it, and then stretch out the leads so that the bridge is from the same row as the bottom of R4 and GND. You can stretch the leads by dragging out from the red area at the end of the leads.
- **LED2:** This is similar to LED1 except that it will go at the top of the board. Rotate it and flip it to get the part the right way around, and then splay out one lead to meet the positive power supply column. The other pin of the LED goes on the same row as the top of R3.

There is one more wire to add because pin 4 of IC1 needs to go to the positive supply, so drag the wire out from pin 4 of IC1's row on the breadboard to the positive supply column.

After all the components have been placed and the labels moved around a bit, the Breadboard view should look like Figure 2-18.

Batteries Not Included

There is a glaring omission from this project: there is nothing to supply it with power. We added all the components so far by using the Schematic view. But there is no reason why we shouldn't add parts from the Breadboard view. To prove that point, you are going to add a battery to power the LED flasher.

Search for "Battery" in the parts window, and drag a 9V battery onto the breadboard. Then drag from the red positive connection of the battery to the positive supply column of the breadboard and the black negative lead to the GND column.

The breadboard should now look like Figure 2-19.

If you switch over to the Schematic view, you will see that the battery has been added with rat's nest wires (Figure 2-20a). Rotate the battery and drag out proper wires to complete the schematic (Figure 2-20b).

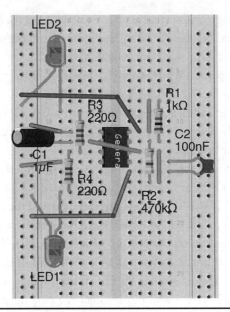

Figure 2-18   All the parts on the breadboard layout.

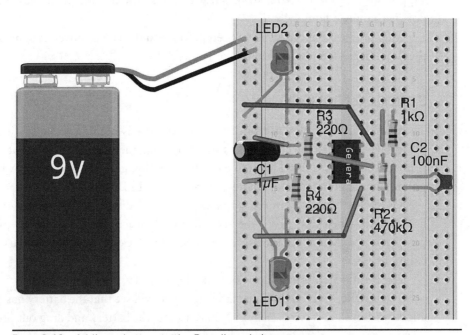

Figure 2-19   Adding a battery to the Breadboard view.

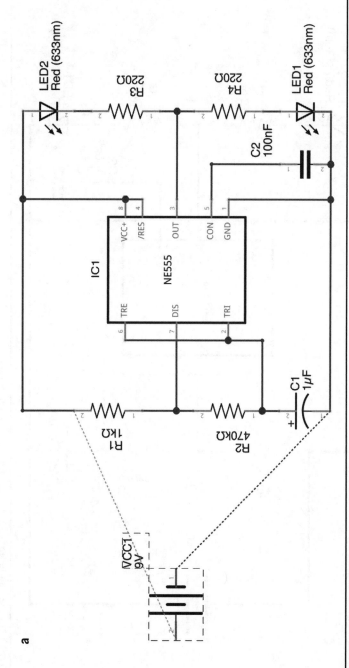

Figure 2-20 Tidying the battery on the Schematic view.

27

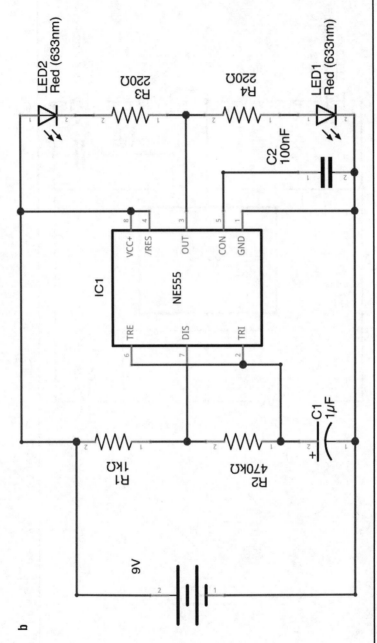

Figure 2-20 Tidying the battery on the Schematic view (*continued*).

28

Build the Breadboard

At this point, you now have a full breadboard design. If this were a project of your own devising, now would be the time to make the breadboard design for real and test it out. We will return to this in Chapter 4.

Designing the PCB

Fritzing will keep all three views (Schematic, Breadboard, and PCB) in step with one another. When you switch to the PCB tab, you should see something like Figure 2-21.

Position the Parts

There are some components on top of one another again, so we can start in the same way as we did with the Breadboard view, by tidying up the components. As with the breadboard, it will make life easier if the components that connect to the pins on the left-hand side of IC1 are toward the left, and those that connect to the right are on the right. It is also a good idea to put the IC near the middle of the board.

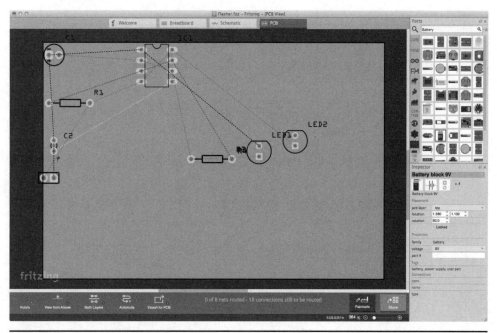

Figure 2-21   Initial PCB layout.

You can make the PCB a bit smaller by dragging the bottom corner, or changing the values of shape in the Inspector. I went for 50mm × 40mm. Move the parts around so that they are roughly in the positions shown in Figure 2-22.

You can see, in the top left, a rectangular part marked with a question mark. What has happened here is that Fritzing has realized that we are not going to solder a 9V battery to the PCB and so replaced the battery with a connector to which we can solder leads or a header connector.

The rat's nest wires now indicate where copper tracks on the PCB should go. PCB manufacturing services all produce double-sided PCBs. That is, you can have copper tracks on both the top and bottom surfaces of the PCB, with "vias" that can connect a track on the top of the board with one on the bottom. Tracks on the top can also be connected to tracks on the bottom wherever there is a hole for a part lead.

Run the Autorouter

In Chapter 6 we devote a lot more time to the whole business of routing a PCB, but for now you can just let Fritzing do the work by running its autorouter. To run the Autorouter, just click on the Autoroute button at the bottom of the window. After a bit of trial and error, Fritzing will have laid out the board, using different colors for the top and bottom layers. The result is shown in Figure 2-23.

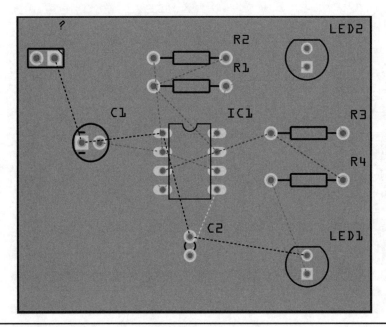

FIGURE 2-22   The parts rearranged on the PCB.

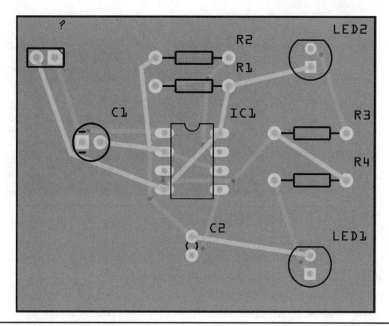

FIGURE 2-23   The autorouted board.

Run the Design Rule Checker

The Autorouter is not guaranteed to lay out your PCB perfectly every time, so whether you lay out the board by hand or use the Autorouter, you must run the design rule checker (DRC) afterward. This tool will check for any tracks crossing that should not or anything that is too close together. To run the DRC, select the menu option Design Rule Check from the Routing menu. This will open up a little window reporting any problems (Figure 2-24).

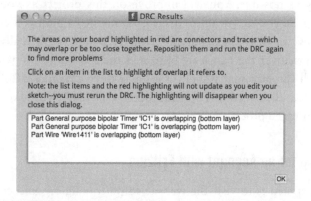

FIGURE 2-24   Results of the design rule check.

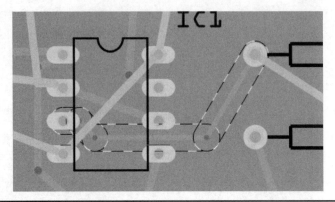

FIGURE 2-25   The modified routing.

When you select one of the problems in the list, the area causing the problem on the PCB will be highlighted with little red marks.

All three errors relate to the track from pin 3 of IC1 being too close to the solder pads for pins 4 and 5 of IC1. We can solve this by rerouting the errant track a little bit. Drag the bend points for the track away from pin 4 of IC1, and then make the track a bit thinner so that it passes between pins 5 and 6 with a bit more room to spare. The width is changed by selecting the track and then changing the width to 16 mils in the Inspector. After you have made these changes, run the DRC again. You should get a message to the effect that the design is ready for production. The modified routing is shown in Figure 2-25.

PCB Manufacturing

This is a very rough-and-ready design, intended to get you up and running as quickly as possible. You will improve on this project as you learn more about Fritzing. So you may want to delay ordering PCBs until you are ready.

However, having Fritzing make the PCBs for you is really easy. If you hover the mouse over the Fabricate button at the bottom of the window, Fritzing will even tell you how much the PCB is going to cost (Figure 2-26).

If you click on the Fabricate button, Fritzing will take you to a webpage describing the process.

Sign Up for an Account with Fritzing

The first step is to visit the website http://fritzing.org and click on the Sign Up button. If you have already signed up, then you can just click on Login. As well as

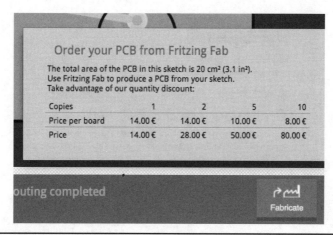

FIGURE 2-26   Fritzing prices for fabrication.

allowing you to use the PCB service, registering with Fritzing will allow you to post on the Forum, where you will find lots of users ready and willing to help you with any problems.

Fritzing is a community project, and if you have a design that you want to share, then you can upload the design to the Fritzing website for others to use and learn from. You will find this option on the Projects link at the top left of the page.

Upload the File and Pay

Once you are registered with Fritzing.org and signed in, click on the Fab (Fabricate) button at the top of the page. This will take you to the fab.fritzing.org website. Click on the Submit Your Order button, and a new page will open, as shown in Figure 2-27. Click on Add Sketch and then navigate to the file Flasher.fzz (make sure you have saved it). Note that *sketch* is Fritzing's word for a design.

Notice that the manufacturing cost has been confirmed here as well as information about when you are going to get your boards.

The rest of the process is the same as you would expect with any online ordering system. You have to provide your contact details and then hand over some money.

Sometime later, the mail carrier will deliver you your board or boards, and you can think about assembling the components onto the board. In Chapter 7, you will learn how to solder components onto the board.

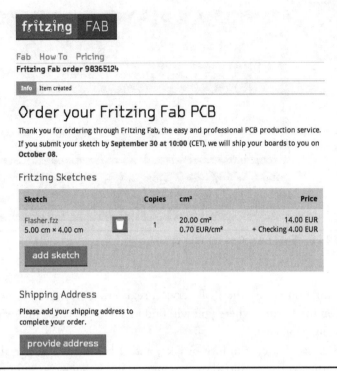

FIGURE 2-27   Adding a sketch to your order.

Summary

This has been a very rapid overview of using Fritzing. I hope it has given you an idea of just how quick and easy this tool is to use. In Chapter 3, you will take a step back and look at the whole process of electronic invention.

Electronic Invention

It is beyond the scope of this book to teach you everything about electronics. For that, you will need a bigger book such as *Hacking Electronics* or *Practical Electronics for Inventors*, both by Simon Monk and published by TAB Books.

At this point, it is useful to look at the various stages that you are likely to go through, when taking your initial idea and converting it to a product. The stages are shown in Figure 3-1. Although these stages will probably all be necessary, perhaps you, as the inventor, only take it so far and then partner with someone who can continue with the final stages of the project.

First, you will have an idea of what you want to make. As an example, let's assume that you want to make a novelty project, a musical instrument that uses an ultrasonic range finder to vary the pitch of a note generated by an Arduino board. This project is based on a project described in the TAB *Book of Arduino Projects*. I will not go into all the details of the whole design process, but rather use it as an example of the kind of things that you need to be thinking about as you design a project.

The Idea

Some people claim to be "ideas" people, painting in broad strokes and not really wishing to understand or be involved in the details, with the latter being a job of lesser mortals (the technicians). Time and again, I have found this to be untrue. The distinction between artist and maker is an artificial one. If you want to invent something, you will find things go much more smoothly if you can make it yourself; even if it is a crude prototype held together with electrical tape and a prayer, you should still be able to make it. All the world's great inventors make

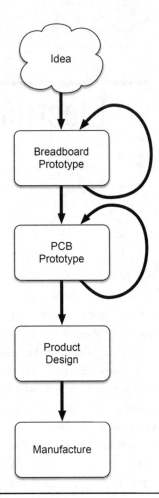

FIGURE 3-1   The product design process.

their own prototypes. Thomas Edison was not just an ideas man. To take a more modern example, the stylish reinventor of the vacuum cleaner James Dyson made prototypes of his cyclone system.

This is not to say that there aren't people (let's call them artists) who have an aesthetic sensibility that allows them to make things look beautiful and stylish. This is a skill much to be admired and somewhat lacking in this author.

Generally, paper and pencil are probably the best tools for sketching out an initial idea. Figure 3-2 shows what this might look like for our musical instrument.

If you are good with a drawing or CAD package, then you may want to use it to describe what you have in mind. Generally, pictures or words on their own are not enough; you will need pictures and accompanying text that explains what your idea does.

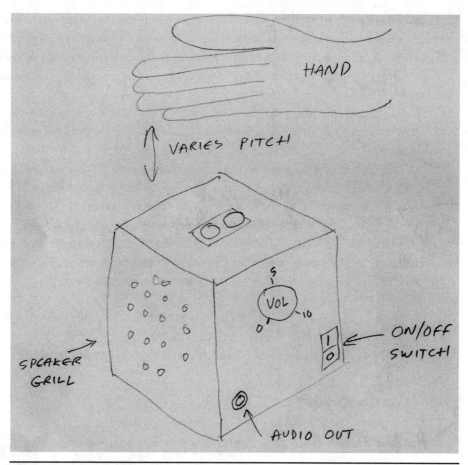

Figure 3-2 A novel musical instrument.

Documentation will be useful not only to you in clarifying your ideas, but also to others to explain how the project will work. It is always worth canvassing other people's ideas. You can ignore their suggestions, but it's still useful to have them.

Do not spend too long specifying what you intend to make. When you actually make it and have a prototype in front of you to play with, you will probably find that your ideas change. So do just enough planning out of the idea to allow you to get on to the next stage of building a breadboard prototype.

Breadboard Prototype

It is a lot easier to use ready-made electronic modules like the Arduino or an ultrasonic range finder module and connect them to a breadboard than it is to design something from scratch. This will get us to the stage of a working prototype, but when you start to look at how you would make lots of these objects, it becomes wasteful and expensive to use modules. For now, while we are proving the concept by making a prototype, the solderless breadboard and ready-made modules are ideal.

In this case, the ready-made modules will be the Arduino, an ultrasonic range finder module, and an audio amplifier module. Figure 3-3 shows the breadboard layout.

Fritzing makes it easy for us to make a nice-looking, easy-to-follow diagram while we are connecting things on the breadboard. In Chapter 4 you will learn a lot more about using breadboards for this stage of the invention process.

Once you have built the breadboard prototype, try out the project and get other people involved in using the device for real. This experience will be invaluable. Make a list of your thoughts and feedback from others. This will allow you to refine your design. This might include such things as these:

- It is too hard to hit just the right note. Could it be made to autotune?
- Could you have a switch to turn autotune on and off?
- It would be good to have another distance sensor to set the volume of the instrument, as on a theremin.

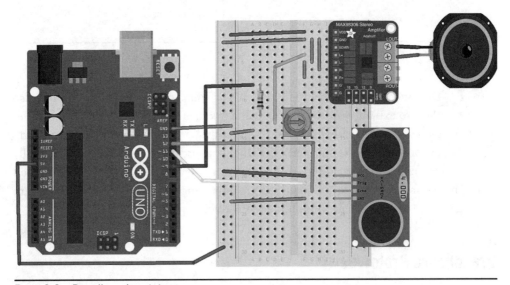

Figure 3-3 Breadboard prototype.

Some of the ideas can be ignored, but others would improve the project. Fortunately, at this stage, you can easily implement the kinds of changes described above. The first one is a matter of changing the Arduino software, but the other two require an extra range finder and slide switch to be added to the breadboard design.

Keep repeating the cycle of prototype and evaluation until you are happy with the design and any software involved in the project. It's worth getting as much mileage out of this design stage as possible because it's so much easier to change things now than in the next stage, which is a PCB prototype.

PCB Prototype

Moving from a breadboard design to a PCB is a big step. An Arduino makes sense for prototyping, because you can simply plug things into it with jumper wires. However, any final product that used the Arduino would be expensive to produce. In actual fact, the main thing that you need from the Arduino is the microcontroller chip (the ATmega328). You can assume that this will be preprogrammed during manufacture or by using an in-circuit serial programmer (ICSP), so you do not need the USB circuitry of the Arduino. You will need to provide some kind of voltage regulator. Thus the basic lesson is that you can look at the open-source schematics for the Arduino and just take the parts of the circuit that you need. The Shrimping.it (see Chapter 8) approach to off-board Arduino design is a great illustration of how this can be quite straightforward. If you are considering an invention with an Arduino at its heart, then you should consider the Shrimping.it approach. The design of Figure 3-3 is repeated in Figure 3-4, but using the Shrimping.it concept.

Design Decisions

These are the kinds of questions you should be asking yourself for this example project:

- What kind of batteries will the final product use?
- Will it have a DC power jack?
- What sort of on/off switch will it have?
- What kind of loudspeaker will it use?
- Will it have an audio output jack?

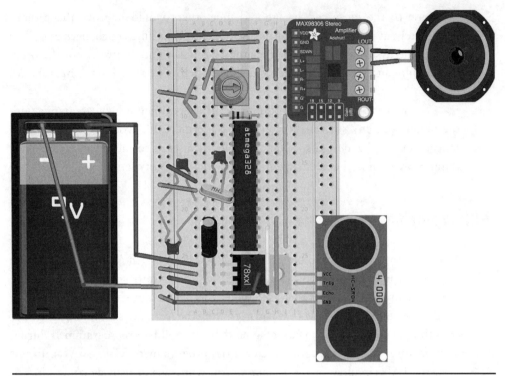

Figure 3-4   Taking the microcontroller off the Arduino.

The fewer external wires the project has, the simpler it will be to make. The speaker will probably be too big to fit comfortably on the PCB, so this will likely be attached with connectors. Connectors will probably also be needed to allow the second range finder to be positioned far enough away from the other range finder. The battery and switch may also be off the PCB and connected via wires, or you may decide to have them both mounted on the PCB.

Cost is also a factor to consider, and you may find yourself spending a lot of time with on-line component catalogs.

A First PCB Prototype

At this stage, you can probably go ahead and design the PCB. In all probability, you will end up making at least one further PCB prototype. So get as small a batch of PCBs made as possible, and solder a board. The Appendix includes links to companies that will make you a set of bare PCBs for as little as $1 per board.

Make sure everything still works, and once again evaluate it and decide what needs to change.

If you are selling a small number of high-value products, then the cost of all the components is not as crucial as if you wanted to sell to a mass market. The cost of the components will be a significant factor in the overall cost of the product. So if you find that one integrated circuit (IC) is particularly expensive, search and you will probably find a cheaper alternative.

Product Design

There is a big difference between a prototype and a product. You only have to make one or two prototypes, and they don't have to be rugged or reliable. When making a prototype, you could buy an off-the-shelf project box and drill holes in the right places for the switches and sockets. Essentially, it does not matter how time-consuming or difficult it is to make. But when it comes to making a product, you have to consider these things in greater detail.

Enclosures

You need to consider the overall shape and aesthetics of the design. If you want something more than a plain rectangular box, you need to design an enclosure for the project. This is a specialized job, but there is no reason why you cannot give it a try yourself, using three-dimensional (3D) modeling software. Use of 3D modeling software is a broad enough topic to merit a book or two of its own; but if you have created the design and you have a 3D printer, you can print it yourself or send the design files away to a 3D printing service.

3D printing is time-consuming and relatively expensive in materials. Once you have a case design that is just how you want it and that matches up with the PCB perfectly, then you can get a quote from a manufacturer to produce the same design by using plastic moldings. This will be a much cheaper cost per unit, but there will probably be a minimum order quantity of at least 500 units.

PCB Manufacture

Once you have made some PCB prototypes by hand, you will probably want to get a quote for the manufacture of these boards in quantity. I have included a list of companies that supply such services in the Appendix. You will also find some links to companies offering 3D design and manufacturing services.

When you submit a PCB design for manufacture and assembly of the components, you need to provide a Bill of Materials (known as a BOM). This may have to fit into the guidelines of the company providing the quote, but will typically consist of a list of all the components including the following information:

- The component name, for example, R1
- The component value, for example, 270Ω ¼W 5% accuracy
- An example source where the component can be bought

There will be other columns on the table such as the cost of the component in various quantities that the manufacturer will complete to estimate the job. The manufacturer may make up one or two of the boards as prototypes for you to test. The manufacturer may also require you to make a test harness to check that the production process is okay. All these things are covered in greater detail later in the book.

Summary

This chapter has summarized the process of creating a new electronic product. In the chapters that follow, we will explore these steps in a much greater detail, starting with the important initial step of creating the breadboard prototype.

CHAPTER 4

Breadboarding

In this chapter we will take a step back from where we ended up in Chapter 2 and look in greater detail at using the solderless breadboard to prototype our electronic inventions.

Electronics is a large and complex discipline; however, in this book, we can at least get you started and point you in the right direction by looking at some of the basics. I will do all I can to introduce this with minimal mathematics and jargon. Electronics engineering as a discipline makes great use of mathematics to model what is going on. This is all fine and good if you have three or four years to study the subject, but is somewhat off-putting if you just have a simple idea that you want to realize. This is more possible than ever, as in recent years making your own electronics has become an ever more popular pastime, and there is a proliferation of tools and electronic modules to help you. In other words, there is no longer the need to make everything from first principles every time or to have a deep understanding of all the technology involved.

Still, there are many good self-taught electronics experts out there who have acquired their knowledge through practical experience rather than formal learning.

How the Breadboard Works

Breadboard, or more accurately *solderless breadboard*, allows you to build electronic circuits without soldering. This makes it much easier to put together an electronic project and then to change things, but is not normally suitable for long-term use as it's far too easy for wires to become accidentally disconnected.

The basic idea is that component leads and "jumper" wires are pushed through from the top of the board and are connected by the metal clips built into the breadboard on its underside.

Figure 4-1 shows a "half breadboard" that is a useful size for most prototyping. Bigger and smaller breadboards are also available, and it can be useful to have a range of boards. What's more, most breadboards are designed to be clipped together to make bigger boards, if you have more complex projects.

The breadboard in Figure 4-1 is organized into four main areas. On both the left and right of the board are vertical rows of pins. All the holes for one column are connected by a long conductive clip on the underside of the breadboard. This is true for the two columns on each side of the board. These connections will not always be used when you are using the breadboard, but are very commonly used as "supply rails." That is, the positive and negative power supply connections for a project typically need to be connected to lots of parts of the circuit, so these columns

Figure 4-1   A half breadboard.

are often color-coded red for positive and blue or black for negative. That way, jumper wires can be used to connect one of the supply rails to a row in the main area of the breadboard. We did this in Chapter 2; take a look at Figure 2-18 for an example of using the supply rails. Many breadboards allow these supply rails to be unclipped from the main body of the breadboard if they are not needed.

The main area of the board is divided into two banks of rows, each of five holes. Each of these rows of five is connected by a clip on the underside of the breadboard. Figure 4-2 shows the underside of a breadboard that has been disassembled so that you can see the backs of the metal clips that sit behind the holes in the plastic fascia of the boards. One of the clips has been removed and is to the side of the breadboard in Figure 4-2.

Breadboards usually come with an adhesive, soft foam backing. This often has a double-sided adhesive so that in addition to stopping the clips from falling out,

Figure 4-2   The underside of a breadboard.

the whole breadboard can be stuck down to a base. Also if you take the adhesive backing off the breadboard, you can unclip the sidepieces of the breadboard containing the supply rails. This allows you to join as many breadboards as you like in a configuration suited to the project you are prototyping.

Components for Breadboarding

Not all components are suitable for use with breadboards. Many components these days are designed to be soldered to the top surface of a PCB, instead of having leads that pass through a hole in a PCB. Such components are called *surface-mount devices* or SMDs. SMDs are not suitable for use with breadboards as they do not have legs long enough to poke through from the top surface of the breadboard. Breadboard requires the use of the more old-fashioned "pin-through-hole," or PTH, devices.

You can still use through-hole components with breadboard to prototype a design that will eventually use SMDs on a PCB, because most SMD devices have PTH counterparts. So you can prototype using PTH devices and then design for PCB using SMDs. You could just stick with PTH devices all the way through to the final product, but through-hole designs cost a lot more to assemble and are bigger, and the components themselves are also usually more expensive than their SMD counterparts. Figure 4-3 shows a range of components, with SMD and through-hole counterparts side by side.

You will find sources for all the components used here in the appendix.

Now let's look at some key components that you are likely to use in your designs and get an idea of how to draw a breadboard diagram in Fritzing and then how to actually build them on breadboard.

Resistors

The most common use of a resistor is to restrict the flow of electric current. In the next section on LEDs, you will see that you need to use a resistor with the LED to prevent too much current from flowing. If left to its own devices, a LED will draw so much current that it burns itself out. You can see a picture of a resistor in Figure 4-3. Its stripes tell you the value of the resistor, that is, how much resistance to the flow of current it offers. The resistance is measured in ohms (Ω), kilohms (kΩ), or megohms (MΩ). Thus 1kΩ = 1000Ω and 1MΩ = 1,000,000Ω.

A resistor of 1Ω will offer very little resistance to the flow of electric current, and lots of current will flow. At the other extreme, a resistor of 1MΩ will allow very little current at all to flow. Not surprisingly, resistors in the middle of the range are most used.

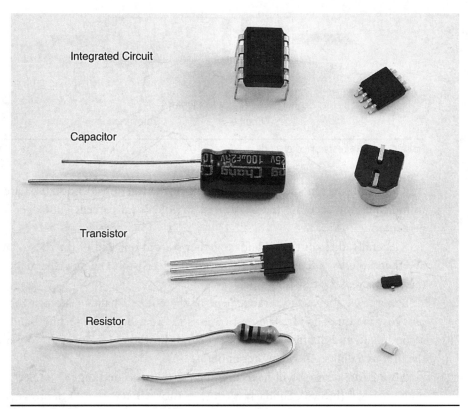

FIGURE 4-3   Through-hole and SMD components.

Resistors have stripes on them that tell you their value. Each color has the value below:

Black	0
Brown	1
Red	2
Orange	3
Yellow	4
Green	5
Blue	6
Violet	7
Gray	8
White	9
Gold	0.1
Silver	0.01

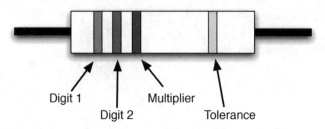

Digit 1   Multiplier

Digit 2   Tolerance

Figure 4-4 The resistor color code.

Gold and silver in addition to representing the fractions 0.1 and 0.01 are used to indicate how accurate the resistor is, so gold is ±5 percent and silver is ±10 percent.

Generally, three of these bands together start at one end of the resistor, a gap, and then a there is a single band at one end of the resistor. The single band indicates the accuracy of the resistor value.

Figure 4-4 shows the arrangement of the colored bands. The resistor value uses just the three bands. The first band is the first digit, the second band is the second digit, and the third "multiplier" band shows how many zeros to put after the first two digits. All values are in ohms.

So a 270Ω resistor will have first digit 2 (red), second digit 7 (violet), and a multiplier of 1 (brown). Similarly a 10kΩ resistor will have bands of brown, black, and orange (1, 0, 000). Some resistors of higher accuracy have four value stripes rather than three. In this case, the first three stripes represent the value number, and the final stripe is the number of zeros on the end.

When resistors restrict the current flow, they get hot. If they get too hot, they burn out. For this reason, resistors are supplied in different power ratings. Generally, the higher the power rating, the bigger the resistor will be physically. Most people use resistors with a ¼W power rating (W for watt). This is fine nearly all the time.

LEDs

LEDs create light when an electric current passes through them. They are fast replacing lightbulbs in all sorts of applications. They can be used as indicator lights, and more recently high-power LEDs are used to provide illumination. They are available in pretty much any color you could want. What's more, you can get LEDs that can change color. These LEDs are called RGB (red, green, blue) LEDs. They are actually three LEDs very close together and within one plastic package that diffuses the light from the red, green, and blue LEDs so that the color is mixed. If the red, green, and blue LEDs are all producing equal amounts of light, then the LED will appear to emit white light.

Basic single-color LEDs are most commonly found as red, yellow, green, and blue, but are also available in more exotic colors and even beyond the range of human vision as Infrared (used in TV remotes) and ultraviolet (used to check bank notes).

All these LEDs are used in much the same way. One lead is longer than the other, and 99.9 percent of the time this is the positive lead, with the shorter lead being the negative lead. Unfortunately you cannot just connect an LED to a power source such as a battery. Well, you can, but at best it will shorten the life of the LED, and at worst it will simply kill the LED. You have to supply power to the LED via a resistor to limit the current flow. You will often see the kind of arrangement shown in Figure 4-5.

To get some practice drawing diagrams with Fritzing, I suggest you first draw the diagram in Fritzing and then, if you have the parts, actually make it up on breadboard. Note that all the Fritzing files for this book are available from the book's webpage at www.simonmonk.org.

If you found the quick-start chapter a little too quick, then fear not! In the sections that follow I will lead you much more slowly through the use of Fritzing to make breadboard designs.

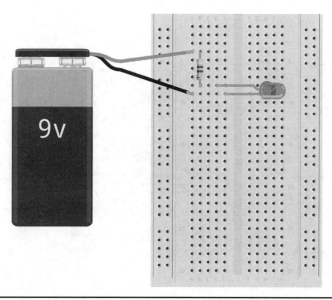

Figure 4-5   A resistor and an LED on breadboard.

Step 1. Change the Breadboard Size

First, start a new Fritzing design by selecting File | New from the menu bar. Initially, you will be greeted with a full-size breadboard. Select the breadboard; then in the properties panel on the right, select half+ as the size (Figure 4-6).

Rotate the breadboard so that it is vertical with the row numbers starting at 1 at the top of the board. To do this, either click on the Rotate button (bottom left of the window) three times, or select Rotate 90 degrees Clockwise in the drop-down menu next to the Rotate button. By default the Rotate button rotates 90° counterclockwise.

Step 2. Add a Resistor and Set Its Value

The most common parts are all contained under the CORE section in the parts area in the top right of the Fritzing window. In fact, a resistor is the first item in this list. Drag it onto the working area, near but not actually onto the breadboard; then rotate it to be vertical in the same way as you did with the breadboard.

While the resistor is still selected, edit its resistance property to make it 470 (Figure 4-7). Note that you do not need to enter the symbol for ohm (Ω). This field is quite smart, so if you type 1k into it, it will recognize this value as being 1kΩ. Notice that when you change the value in this field, the color of the resistor's stripes will automatically change to match the new value.

Now move the resistor onto the breadboard at the position shown in Figure 4-7.

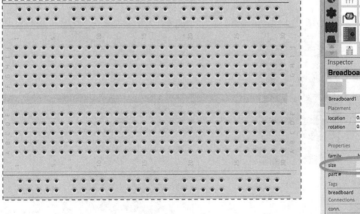

FIGURE 4-6   Changing the breadboard size.

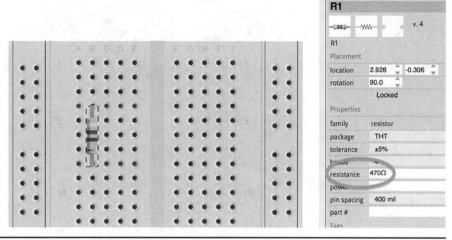

FIGURE 4-7   Adding a resistor.

Step 3. Add an LED

You will also find a red LED in the CORE parts section. Drag it onto the breadboard at the position shown in Figure 4-8. If you hover the mouse over the tips of its two leads, a little notification will pop up identifying the leads as being the anode and cathode, respectively. The *anode* is the positive lead, and the *cathode* is the negative lead. The positive lead also has a bend in it near the LED body. At this point, I find it helpful to extend the positive lead by one breadboard hole in distance by dragging the red tip of the lead down a little.

By convention, breadboards tend to be laid out so that the positive voltages are toward the top of the board and the ground or negative supply is toward the bottom of the board. This allows you to visualize the current flowing from the positive supply to the negative from top to bottom. To get the LED in the right orientation so that the positive lead is toward the top and the LED body is facing toward the right, click Rotate once and Flip once. You can then position the LED on the breadboard next to the resistor, as shown in Figure 4-8.

When you position both the resistor and the LED, make sure that the row of the breadboard that the pins are connected to turns green to indicate a good connection.

Step 4. Add a Battery

We need to supply power to the LED, so we will use a 9V battery. There isn't a battery in the CORE section of the components, so search for it by clicking on the magnifying glass icon in the parts area and then typing **9V battery**. This should

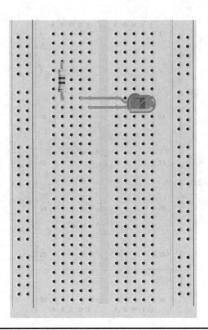

FIGURE 4-8   Adding an LED.

bring back a couple of choices. Pick the one shown in Figure 4-9, the one without the black background.

Position the battery to the left of the breadboard.

Step 5. Connect the Battery

Connect the positive side of the battery to the top of the resistor by dragging from the connection point on the positive battery terminal to the hole on the breadboard immediately to the left of the resistor lead. You may want to click on the background on the way to bend the lead, in order to make the diagram clearer. When the wire has been drawn, select it and change its color to red in the Inspector.

Do the same with the negative lead, but connect it to the bottom negative lead of the LED and change the wire's color to blue. When this is complete, the diagram should look like Figure 4-9.

Let's analyze what we have just done.

As you can see, the positive lead of the battery is connected to one end of the resistor because they are both on the same row. The other end of the resistor is connected to the positive end of the LED (the longer lead), and the negative lead of the LED returns to the negative connection of the battery. You can see now why people refer to electronic *circuits*. The electricity is flowing in a loop.

The current flowing through the LED and resistor is measured in units of amperes, abbreviated to A. An ampere is quite a lot of current so the unit

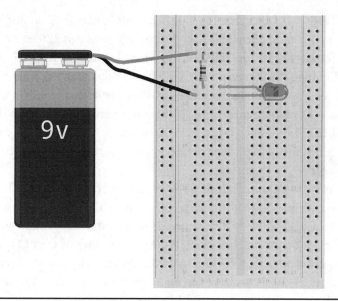

Figure 4-9   An LED and resistor wired up.

milliamperes, abbreviated as mA (0.001A), is more common for small electronic projects. Most standard LEDs are designed to give an optimum brightness without any chance of damage to the LED when a current of 20mA flows through them. They will actually produce some light with as little as 1mA, and you will not be able to see much difference in brightness between 10 and 20mA. In other words, the current flowing through them can vary over quite a wide range without any problems, which in turn implies correctly that the choice of current-limiting resistor is not critical.

Table 4-1 will help you choose a resistor value, but if you want to know the theory, then take a look at http://en.wikipedia.org/wiki/LED_circuit. There are also online calculators that will do the math for you, such as this one: http://led .linear1.org/1led.wiz.

Table 4-1   LED Series Resistors

Supply Voltage	LED Color		
	Red	Orange, Green	Blue, White
3V*	68–270Ω	47–270Ω	None
5V, 6V	220–470Ω	180–470Ω	150–470Ω
9V	470Ω–1kΩ	390Ω–1kΩ	330Ω–1kΩ
12V	560Ω–1kΩ	560Ω–1kΩ	470Ω–1kΩ

* When using a Raspberry Pi or BeagleBone, use a 470Ω resistor for all colors as these devices have very low current outputs.

Table 4-1 shows that for blue and white LEDs, you do not need a resistor at all when you use them at 3V. At 5 or 6V, a resistor of 220 or 270Ω will work just fine for any LED color; at 9V and above, a 1kΩ resistor will do in all circumstances.

Limiting the current to the LED doesn't just stop the LED from burning out; it also prevents damage to whatever is supplying power to the LED. If it happens to be a digital output from an Arduino or microcontroller chip that is controlling the LED, then those outputs are designed to supply only a certain amount of current. If that current is exceeded, then the Arduino or microcontroller might be damaged.

As we mentioned before, an RGB LED is actually three LEDs in one package. But rather than have all six pins emerge from the package, most RGB LEDs have just four pins. There will be a pin for each of the three colors and a common pin. The common pin will be either all the positive ends of the LEDs connected (called *common anode*) or all the negative ends of the LEDs connected (*common cathode*).

Figure 4-10 shows how you might wire an RGB LED. You will need to use the search facility to find the RGB LED. Just for practice, draw the breadboard design; then if you have the parts and feel like it, make the circuit for real.

With all the resistors plugged in, the LED should emit a white light. Try unplugging one end of each resistor and notice how the color changes.

FIGURE 4-10 Using an RGB LED with breadboard.

Switches

Switches connect two metal contacts when you press them or slide them, depending on the type of switch. You can divide switches into two categories: those that connect the contacts only while they are being pressed and those that once put into a "closed" position (contacts connected) stay connected until someone flips the switch the other way. I start by considering the push switch.

You can easily adapt the circuit of Figure 4-9 to include a push switch so that the LED does not light until the button is pressed. Figure 4-11 shows the modified breadboard layout.

These switches can be confusing to use because they have four pins where you would expect them to have only two. The extra two pins are only really there as a by-product of the switch construction and to give them a good solid fix onto a PCB, so that they can be pushed hard with no chance of their moving. With the switch in the position shown in Figure 4-11, the switch contacts on the same rows are connected internally. That is, you can just use the pins on the left of the switch as shown and ignore the other two.

Figure 4-12 shows a slide switch.

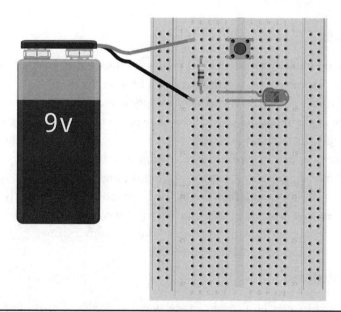

FIGURE 4-11 Using a push switch with breadboard.

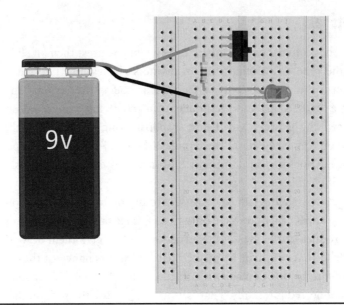

Figure **4-12**   Using a slide switch with breadboard.

This type of switch has three connections. The middle pin of the switch is connected to the top pin or to the bottom pin depending on the switch position. With the switch lever in the position shown in Figure, 4-12, the positive lead of the battery will be connected to the top of the resistor and the LED will light.

Diodes

You are unlikely to use normal diodes (as opposed to light-emitting diodes) very much. However, it is worth knowing a little about them.

Diodes allow current to flow in only one direction. They are like a one-way valve in plumbing and so can be used to protect a circuit against accidental reverse voltages. In fact, this is exactly what you will use one for in the first example of the next section, where a motor is being switched on and off by a transistor.

Since diodes allow current to flow through them in only one direction, it is obviously important that you have them positioned correctly in your circuit. Diodes have a stripe at one end, so you know which end is which. The current flows from the end with no stripe to the end with a stripe.

Transistors

Transistors can be used in many ways in analog electronics, for example, to amplify signals. However, nowadays integrated circuits will usually fill these roles rather than single transistors. However, transistors are very commonly used as electronic switches, and it is this use of transistors that we concentrate on here.

In an earlier section on using an LED, we said that limiting the current to an LED protects both the LED and whatever is supplying the current. Microcontrollers and microcontroller prototyping boards such as the Arduino have digital outputs that can typically supply a maximum of around 40mA. That's more than enough for a small LED. But what if you want your microcontroller to turn much more powerful things on and off? Examples might include motors (which usually use lots of current) or LED lamps used for illumination rather than as indicators. It is quite common to need to control something that might require hundreds of milliamperes or even several amperes. In this section, we look at two common transistors: the 2N3904 that will switch up to 100mA and the bigger FQP30N06 that can switch up to 32A at 60V (with the right heat sink attached).

Transistors have three pins. The names of these pins vary depending on the type of transistor, which is confusing, but you can think of them as current in, current out, and a control pin. Figure 4-13 shows how you can use a 2N3904 transistor to turn a small DC motor on and off.

When you draw this diagram in Fritzing, you will not find a 2N3904 transistor in the parts bin. Instead, the parts bin has some standard transistor types near the top of the CORE parts. You want the one shown in Figure 4-13 with an N in the transistor body. Technically speaking, this type of transistor is an NPN (negative-positive-negative) bipolar transistor. Before you add it to the breadboard diagram, drag the transistor legs back toward the body of the transistor to make them as

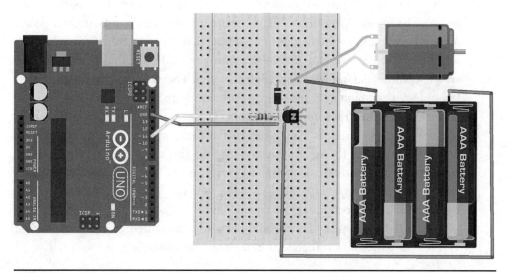

Figure 4-13   Controlling a small DC motor with a 2N3904 transistor.

small as possible. If you don't do this, they will cover most of the breadboard row, leaving no room for connections and other components. Having selected the transistor, you can change the "part#" field so that you know what type of transistor you used.

The Arduino pin 11 is connected to the middle connection of the transistor (the base) via a 470Ω resistor. As with an LED, you need to limit the current flowing into the base of the transistor with a resistor. A value of 470Ω will suit most situations. When this Arduino output is set high (to 5V), the transistor will turn on and allow current to flow through the motor, making it spin.

The bottom connection of the transistor (the emitter) is connected to the Arduino GND connection and to the negative connection of the battery that will supply power to the motor.

Note that this will not power the Arduino; you will need to power that by other means.

One end of the motor is connected to the positive battery connection, and the other end goes to the top connection of the transistor (the collector).

If we were driving, say, a high-power LED, then that would be all there was to the circuit. However, motors and other inductive loads such as a relay (anything with a coil of wire in it is an inductive load) can produce spikes of voltage in the opposite direction to the direction it went in when power is removed. The diode protects against this. Because the diode stripe is at the positive end of the battery, normally the diode does not conduct. However, if there is such a negative spike of voltage from the motor, the diode will short it out, preventing it from doing any damage.

If you need to control more than 100mA, then a 2N3904 transistor is not going to be enough. You need a bigger transistor that can handle greater power. A popular choice for this is the FQP30N06. This type of transistor is called a *metal-oxide semiconductor field effect transistor* (MOSFET). These transistors are great for switching relatively high currents. They are built using a different technology than bipolar transistors, and so while they still have three pins, the pins are called the *drain*, *source*, and *gate*. The gate controls the flow of current like the base, the drain is connected to the more positive end of whatever is being switched, and the source goes to ground. This naming convention is quite counterintuitive and stems from an early error in the understanding of electronics. If you are interested in why, you can find an explanation here: www.allaboutcircuits.com/vol_1/chpt_1/7.html.

Figure 4-14 shows the FQP30N06 being used to switch what we assume will be a higher-power motor than that in the previous example.

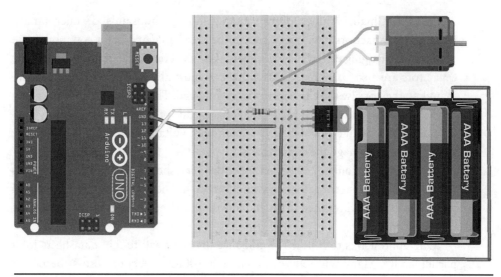

Figure 4-14   Controlling a DC motor with a MOSFET.

The first thing to notice is that the diode has gone. This is so because MOSFETs have built-in protection diodes. The other thing of note is that the pins of the MOSFET are not in the same order. The controlling pin, the gate, is now at the top rather than in the middle.

When looking for the MOSFET in the parts bin, you should select "Basic FET" from the CORE section and then select "n-channel" for the transistor's type in the Inspector. The MOSFET will quite happily control 4 or 5A of load as it is. For higher currents you will need to bolt it to a metal heat sink.

Capacitors

Capacitors are components that store small amounts of electricity. In that sense, they are a bit like batteries, except that size for size they do not store nearly as much electricity as a battery, and then they can fill (charge) and empty (discharge) extremely quickly.

The unit of capacitance (a measure of how much charge they can hold) is the farad, abbreviated as F. Most capacitors are in the nanofarad (nF) range (1/1,000,000,000F) or the microfarad (µF) range (1/1,000,000F).

The most common types of capacitor are multilayer ceramic, often just called ceramic capacitors and electrolytic capacitors. Ceramic capacitors are used for lower-value capacitors, generally less than 1µF although ceramic capacitors are now becoming available in ranges up to tens of microfarads. Electrolytic capacitors are used for higher values of capacitance up to thousands of microfarads. Electrolytic capacitors have a tendency to be unreliable and are a common cause of problems in aging electronic equipment. Unlike ceramic capacitors, electrolytic capacitors are also polarized. That is, one end must always be kept more positive than the other, or else the capacitor can be damaged. The negative lead is normally marked as such, and as with LEDs, the longer lead is the positive lead.

Capacitors of all types also have a maximum voltage rating. Again, exceeding this is likely to damage the capacitor.

In digital electronics, capacitors meet the very short-lived but high demands that often arise when transistors within ICs switch. They improve the reliability of your circuits. Knowing when to use them without a deep knowledge of engineering can be a problem.

One approach is to follow the guidance that you will find in datasheets for a particular IC. Many will tell you to place a 100nF capacitor across the power supply pins of a chip as close as possible to that chip. Voltage regulator chips will often require electrolytic capacitors to act as reservoir capacitors to maintain stability.

When adding capacitors to a design in Fritzing, you will find them on the top row of the CORE parts bin category. Here you will find both ceramic and electrolytic.

Fritzing Tricks

While we are looking at creating breadboard designs in Fritzing, it makes sense to cover some of the options available to us in Fritzing to improve our diagrams.

Labels

When a design only has a few components, you can get away without labels. Behind the scenes, whenever you add a new part to your design, Fritzing will give it a name. You can see that name and also edit it in the Inspector area of the Fritzing window. Fritzing will allow you to change one part and give it the name of an existing part, without any warning message. So be careful to use a name that is not already used, if you are using your own names for parts.

The naming convention for parts starts with a letter, with R for resistor, C for capacitor, D for a diode, and for some reason Q for a transistor. This letter is then followed by a number.

FIGURE 4-15   The parts context menu.

By default, the parts labels are not shown of the Breadboard view. You can turn this on and off for a particular part by right-clicking on it and selecting the option "Show part label" in the drop-down menu that appears (Figure 4-15).

Take a look at some of the other options on this menu, because they can be useful. The Rotate submenu can be a more convenient way of rotating a part than using the Rotate button at the bottom of the Fritzing screen. The Raise and Lower submenu helps you to make your document clearer by controlling which parts appear on top of which other parts. This menu also has options for copying and duplicating parts.

Parts Bin

It can sometimes be tricky to find things in the parts bin. If you enable the Show Parts Bin List View option from the View menu, the icon view of the parts will change to a list view that I find easier to use.

Summary

In this chapter, you have the briefest overview of some of the main electronic components that you are likely to use with your breadboard. As you progress through the book and look at other examples, more components will be introduced and a little more electronics theory will be gently wafted in your direction.

In Chapter 5 you will learn about schematic diagrams, how to understand them, and how to draw them in Fritzing.

Schematic Design

When designing simple projects, or projects that simply add inputs and outputs to a mirocontroller, you can often work directly from the breadboard design. If you work like this, you probably want to stay away from the Schematic tab, because if you look at it, it will probably be a mess with components all over the place.

As soon as projects become just a little bit complex, or you want to share your design with someone, then an easy-to-read schematic is necessary. Once you are familiar with schematic diagrams, you may find it easier to start with a schematic and then move on to the Breadboard view.

Reading Schematic Diagrams

Before we can reasonably expect to draw schematic diagrams, we need to be able to read them and at least get the gist of what they are saying. Although schematics look difficult to understand, there are conventions that make them easier to read.

The lines on a schematic represent an electrical connection, be it a wire or a clip on the breadboard or a copper track on the surface of the PCB. These lines link the components that make up the circuit.

Component Symbols

Component symbols don't generally look like the actual components. They are symbolic representations of the components and therefore say more about how the components operate than what they look like. The resistor symbol is a good case in point. Its zigzag line looks like something that might resist the flow of current.

Table 5-1 shows some of the most common components.

TABLE 5-1   Common Schematic Symbols

Symbol	Component	Use
R1 220Ω	Resistor	Reducing the flow of current
C1 100nF	Capacitor (nonpolarized)	Temporary charge storage
C2 1μF	Capacitor (polarized)	
Q1	Transistor (bipolar NPN, e.g., 2N3904)	Medium current switch
Q2	Transistor (MOSFET N-channel)	High current switch
D1 1N4001	Diode	Prevents current flow in the wrong direction
LED1	Light-emitting diode (LED)	Indication and illumination
Battery 9V	Battery	Power supply
S2	Switch	Turning things on and off

In this book, we will use the same circuit symbols as in Fritzing. These are based on the U.S. standard of component symbols. Europe has a slightly different set of symbols, the main difference being that in the European system, resistors are represented as rectangles rather than zigzag lines.

Schematic Versus Breadboard

Figure 5-1 shows the schematic diagram (Figure 5-1a) and the breadboard layout for the same project (Figure 5-1b) side by side, so that you can compare the two representations.

In this case, relating the two diagrams is easy, because the components are more or less in the same place.

Conventions

In theory, you can place the components in a schematic in any crazy positions you like and connect them by the most bizarre connections, and it will still be a perfectly valid way of drawing the circuit. Figure 5-2 shows an example of such madness.

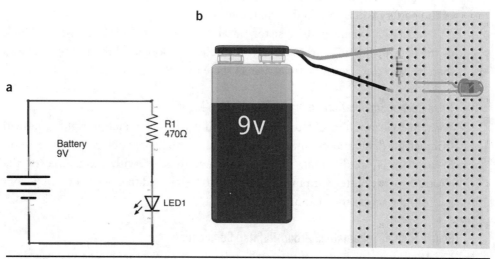

Figure 5-1   Schematic and breadboard diagrams.

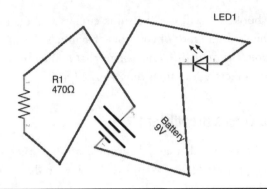

Figure 5-2   A mad schematic diagram.

First Rule of Schematics: Straight Lines at Right Angles

Electronics engineers tend to be quite tidy in their work. Schematic diagrams are drawn with straight lines at right angles to one another. Lines are allowed to cross one another on the diagram, but if the lines are to be joined together where they cross, then a dot is placed over the junction of the wires.

Second Rule of Schematics: Positive Voltages Uppermost

It is a convention that most people follow, when they draw a schematic, to put the higher voltages near the top, so that on the left-hand side of the diagram, we have a 9V battery. The bottom of the battery is at 0V or GND or Ground, and the top of the battery will by 9V higher than that.

Notice that we draw the resistor R1 above the LED (LED1), so we can think of the current flowing through the resistor, before flowing through the LED and then back to the negative connection of the battery.

Third Rule of Schematics: Things Happen Left to Right

Western civilization invented electronics, and it writes from left to right. You read from left to right, and culturally speaking, more things happen from left to right. Electronics is no different, so it is common to start with the source of the electricity—the battery or power supply on the left—and then work our way from left to right across the diagram.

Fourth Rule of Schematics: Group-Related Components

Keep components that are functionally related to one another close to one another. For example, keep the current-limiting resistor for an LED close to the LED.

Names and Values

It is a good idea to give every component in a schematic a name. So, in this case, the battery pack is called just Battery, the resistor is R1, and the LED is LED1. This means that when you go from a schematic to a breadboard layout and eventually a circuit board, then you can see which components on the schematic correspond to which components on the breadboard or circuit board.

It is also usual to specify the value of each of the components, where appropriate. So, for example, the resistor's value of 470Ω is marked on the diagram.

Tidying a Schematic in Fritzing

If you followed my advice in Chapter 4 and drew the breadboard layout shown in Figure 4-12, you should now have a Fritzing project with a breadboard layout that looks like Figure 5-3a; and if you switch to the Schematic tab on that project, it should look something like Figure 5-3b.

The layout of Figure 5-3b is the default layout made by Fritzing on your behalf as you drew the breadboard layout. It's a bit of a mess, but that's okay because you will now tidy up this schematic one step at a time.

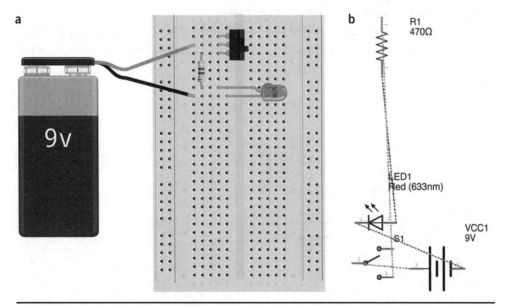

Figure 5-3 Starting point for schematic tidying.

Step 1. Position the Components

I like to start by moving the components around, so that they follow the left-to-right and higher voltages at the top rules that we discussed earlier. It's easier to get things lined up and your lines at right angles if you turn on Align to Grid in the View menu. Start by rearranging the components as shown in Figure 5-4.

Move the battery over to the left, and rotate it so the positive connection is at the top. Next, position the switch, the resistor, and the LED, rotating where necessary. At this point, you can also right-click on the LED and deselect the option to display the LED's color attribute. Also double-click the battery's current name (VCC1) and rename it Battery.

You can rotate the text of the labels independently of the component as a whole by using the menu that pops up when you right-click on the label. You will need to do this for the battery and the LED.

You can also move the labels around relative to the position of the component just by selecting the label and then dragging it.

Step 2. Routing the Nets

Each of the lines on the schematic is called a *net*. In fact, were there any places in this schematic where there were lines that were connected, they would all be part of one net. Initially, in the Schematic view, the component leads are connected with dashed lines to indicate that these lines have not yet been routed. That is, we have not drawn proper lines between them.

Although you can ask Fritzing to route the schematic for you by using the Autoroute option on the Routing menu, it does not always work perfectly, so it's better to do this task by hand.

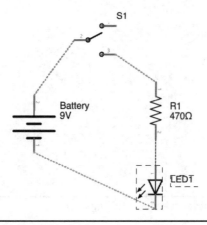

FIGURE 5-4   Components in the right place.

The easiest way to do this is to drag a line out from one of the leads to its connection to another component along the route you want it to take. Figure 5-5a shows this for the connection between the negative battery lead and the LED. When you let go of the mouse button, the line will appear between the two component leads. You can also add a bend point in the line by double-clicking on the line and then dragging the bend point where you want it.

If the dashed line that needs turning into a wire is just a straight line, then double-clicking on the dashed line will convert it.

When all the lines are added, the schematic will look like Figure 5-5b.

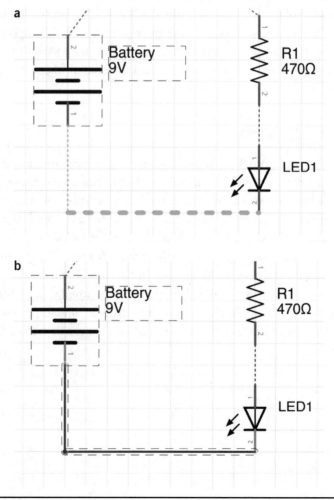

FIGURE 5-5 Routing a net.

Creating a Schematic in Fritzing

In the situation where you are starting with a blank schematic, creating the schematic involves most of the same skills that you use when you are tidying a schematic created from a breadboard design. There is, however, an additional first step of adding the components that you need.

In Chapter 2, we briefly used an example of a 555 timer project to illustrate drawing a schematic. In this example we will take that initial design and modify it to use multiple flashing red LEDs as a rear cycle lamp. The basic design is much the same as that of project 2, so we will begin with a schematic design and then move on to the PCB design for this project in Chapter 6.

If you followed along with the quick-start chapter, you should find it easy to follow these instructions to create the schematic for this project. The end result that you are aiming for is shown in Figure 5-6. You may wish to create this without following the step-by-step instructions, but revert to the instructions if things go wrong.

Design Calculations

Before we wade into the schematic design proper, there are a few design calculations that we need to make. If you are completely new to electronics, then these may seem a little advanced. Since this book is primarily about using Fritzing rather than electronic design, you may find it useful to have a look at a basic electronics text such as *Hacking Electronics* or *Make: Electronics*.

It would be very nice to power this project from the cheap and readily available AAA batteries. These are small, but not so small that they won't last a decent while. You can do a rough calculation of how long a battery will last by using Table 5-2 to look up the capacity of the battery in milliampere-hours (abbreviated as mAh). That figure is the number of hours that the battery can supply a current of 1mA.

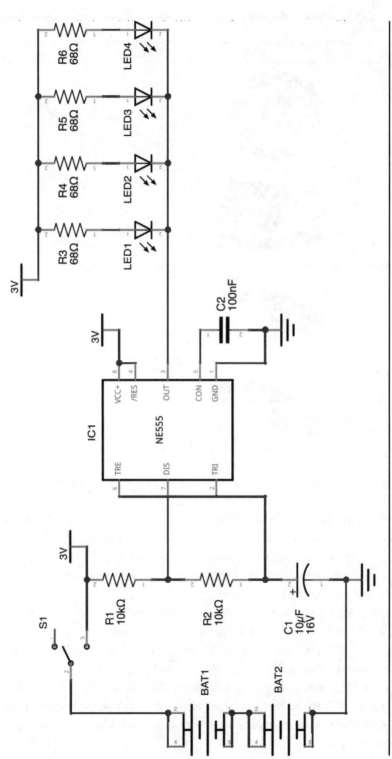

Figure 5-6 Schematic for rear bicycle light.

71

TABLE 5-2   Battery Options

Type		Typical Capacity	Voltage
Lithium button cell, e.g., CR2032		200mAh	3V
Alkaline PP3 battery		500mAh	9V
AAA cell		800mAh	1.5V
AA cell		3000mAh	1.5V

From this you can see that a AAA battery can supply 1mA for approximately 800 hours. This is very approximate, as there are no guarantees for batteries, especially if they have been on the shelf for a while. Now, we have 4 LEDs, and if we assume 20mA per LED, that makes 80mA. But then the LEDs are also flashing, so if we assume they are on only one-half of the time, then there will be an average current draw of around 40mA. The 555 timer chip will use a pretty small amount of current too, but let's ignore that for now.

So now our AAA battery will last 800/40 = 20h. This is a pretty respectable amount of time. A single AAA battery will only generate a voltage of around 1.5V, but the 555 timer will not operate on such a low voltage, so we will have two batteries one after the other (called *in series*). This will give a total voltage of 3V.

In fact, even 3V is not enough for a standard 555 timer (the NE555) that requires at least 4.5V. Fortunately for us there is a variation of the 555 timer called the TLC555 that operates in exactly the same way as the 555 timer, but only needs 2V or more to operate.

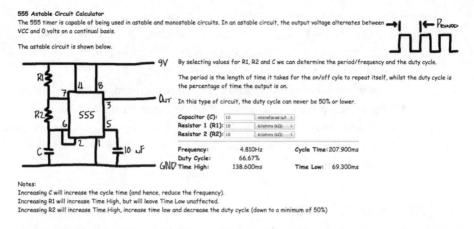

555 Astable Circuit Calculator

The 555 timer is capable of being used in astable and monostable circuits. In an astable circuit, the output voltage alternates between VCC and 0 volts on a continual basis.

The astable circuit is shown below.

By selecting values for R1, R2 and C we can determine the period/frequency and the duty cycle.

The period is the length of time it takes for the on/off cyle to repeat itself, whilst the duty cycle is the percentage of time the output is on.

In this type of circuit, the duty cycle can never be 50% or lower.

Capacitor (C):	10	microFarad (µF ⬍
Resistor 1 (R1):	10	kilohms (kΩ) ⬍
Resistor 2 (R2):	10	kilohms (kΩ) ⬍

Frequency:	4.810Hz	Cycle Time:	207.900ms
Duty Cycle:	66.67%		
Time High:	138.600ms	Time Low:	69.300ms

Notes:
Increasing C will increase the cycle time (and hence, reduce the frequency).
Increasing R1 will increase Time High, but will leave Time Low unaffected.
Increasing R2 will increase Time High, increase time low and decrease the duty cycle (down to a minimum of 50%)

FIGURE 5-7   A 555 timer calculator.

When you used the 555 timer back in Chapter 2, there was no discussion about how to control the speed at which the LEDs flashed. In fact, the frequency of flashing is determined by the values of R1, R2, and C1 in Figure 5-6. There is a formula for this, which you will find if you look at the datasheet for the 555 timer, but it's a lot easier to use an online calculator such as the one at www.ohmslawcalculator.com/555_astable.php. This is also shown in Figure 5-7.

Here C1 is just called C, and there are fields where you can enter values for C, R1, and R2. When you do so, the fields underneath these three fields automatically update themselves. So you can see that with a C value of 10µF and resistors both of 10kΩ, the frequency will be 4.8Hz. That means almost 5 flashes per second. This is probably about the right flash rate.

The Duty Cycle field is underneath the Frequency field. This indicates what proportion of the time the 555 timer's output is high. This is not automatically 50 percent and allows you to make short-duration pulses rather than even flashing if you want. The Time High and Time Low figures express this same information in a different way. If you look at Figure 5-6, you can see that LEDs have one end connected to the positive battery supply via current-limiting resistors. This means that the 555 timer will turn on the LEDs when its output goes low and the voltage flows from the higher voltage to the lower. The LEDs will therefore be on for 69.3ms of the total 207.9ms for each cycle. This means that the LED will be off slightly longer than it is on for each cycle (on one-third of the time). The advantage of this is that our battery will last a bit longer than our earlier calculation would suggest.

It is quite fun to try a bit of trial and error with the component values and see what effect this has on the operation of the project. You should, however, bear in mind that not every single possible value of resistor or capacitor is available. Nor are the actual values exact. For instance, capacitors often only have an accuracy of 10 or 20 percent, so your flashes may vary a little from one build of the project to the next. In this case, that does not matter at all.

Common values of resistors and capacitors are listed in the Appendix.

The final component calculation that we need to make is the value of the current-limiting resistors for each of the LEDs. For this you can refer to Table 4-1. We want the LEDs to be as bright as possible, so let's use 68Ω resistors, equivalent to a current of 20mA.

Drawing the Schematic

The following sections will lead you through the drawing of the schematic one step at a time. To gain practice drawing schematics, follow along; or if you get fed up with that, the Fritzing file for this project can be found in the file cycle_lamp_example.fzz along with all the other Fritzing projects used in this book. These can all be downloaded from the author's website at simonmonk.org. Just follow the links for this book.

Step 1. Add the 555 Timer

If a design is centered about a chip, as it is in this case, then I like to start by positioning it in the center of the design.

Switch to the Schematic tab and use the search facility in the parts bin. Use 555 as the search term. This will bring a few results. Look for the one shown in Figure 5-8. This one is from Sparkfun and names the pins rather than just identifying them by pin number, so it makes for a more meaningful schematic.

Note that the part is labeled NE555. Earlier in this chapter, it was decided that it would be better to use the TLC555 variant of the 555 timer, because of its lower voltage compatibility. There is no TLC555 part in the parts bin, so we will use an NE555 instead. The label NE555 is actually part of the image file used with the part, so it is not so easily changed. However, in Chapter 9, I will show you how to modify and create new parts, so that we can have the right label on our schematic.

Step 2. Add the Timing Components

Having put the IC down, you should start by adding the timing components R1, R2, and C1 to the schematic. These can all be found in the CORE section of the parts bin. The capacitor is of the electrolytic variety. Drag the three components onto the schematic diagram, and rotate the two resistors by 90°. The capacitor is already in the correct orientation.

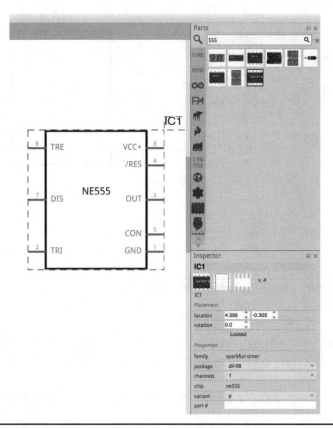

FIGURE 5-8   Adding a 555 timer part to the schematic.

Once the components are in roughly the right positions, set the component values. Both resistors should be given values of 10kΩ using the resistance field in the Inspector. The capacitor actually has two values that need to be set. The capacitance should be set to 10μF and the voltage to 16V. Use the drop-down lists to select these values. The drop-down lists are populated with standard values for these components, so if you are not sure about what values are common, you can use the drop-down lists to allow Fritzing to tell you.

The capacitor has two values: the capacitance, which is obvious, but it also has a voltage value. This value is the maximum voltage that the capacitor can cope with. If this voltage is exceeded, then the capacitor is likely to fail. So generally it's a good idea to err on the side of caution and in this case select 16V from the drop-down list of standard voltage values. In actual fact, the lowest number in the drop-down list (6.3V) would be fine too.

Rotate the labels of the two resistors by 90° clockwise, and then drag all three labels neatly to the left of their respective components. To rotate the label of a

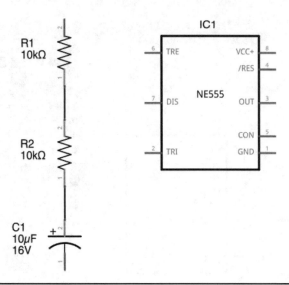

Figure 5-9   Adding the timing components.

component without rotating the whole component, first select the component and then right-click over that component's label, select Flip/Rotate | 90 Degrees Clockwise.

When all this is complete, the schematic will look something like Figure 5-9.

Step 3. Connecting the Timing Components

Drag the connector for the bottom lead of R1 to the top lead of R2. When they connect properly, both connectors will turn green. Connect R2 and C1 in the same way.

Each of these lines connecting components is called a *net*. Even if it's just a single line, it's still called a net. We need to add a connection between pin 7 (DIS) of the IC1 timer to the net that we have just made between R1 and R2. If you try to drag out from pin 7 of the timer IC, then Fritzing will ignore the line that you want to connect to. It will let you connect to either end of the line, but really we want to connect right in the middle of the line.

The trick to doing this is illustrated in Figure 5-10.

First, add a bend point to the line roughly where you want to join the net (Figure 5-10a) by double-clicking on the line. Then drag out from the DIS pin of IC1 to the new bend point. This is likely to be a bit crooked, so finally grab the bend point and drag it so that it is square (Figure 5-10c). This should be done with the Align to Grid feature turned on (on the View menu). Notice how there is a big dot over the

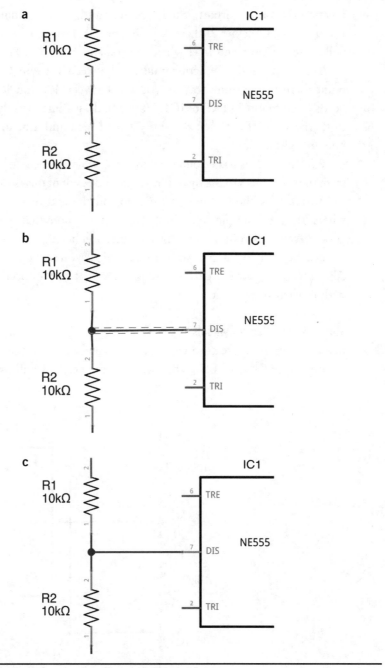

FIGURE 5-10 Adding a connection to an existing net.

junction. This is important; it tells you that all the lines leading off from that dot belong to the same net. If there is no big dot, then there is no connection and this will cause you problems later when you come to make a PCB layout.

You need another junction midway between R2 and C1, so make a bend point there in the same way as you did between R1 and R2. This connection needs to go to pin 2 (TRI) of IC1. Here you drag a line directly from pin 2 of IC1 to the bend point between R2 and C1, and then add a bend point to make the lines at right angles.

Pin 6 also needs to be connected to pin 2, which you can do by simply dragging from the tip of pin 6 to the tip of pin 2. The end result of this is shown in Figure 5-11.

Notice how there is now a little dot on the end of pin 2, but importantly, where the line from pin 6 to pin 2 crosses pin 7, note that there is no dot. That's good, because there should not be a connection here.

A good trick to reassure yourself that things you think are connected are really connected is to move a part and make sure that the connection lines to it move with it when you drag it.

Step 4. Adding Power Symbols

In any schematic, there are always two special nets. One is the ground net. Ground or sometimes GND or 0V is the baseline voltage. This will usually be connected

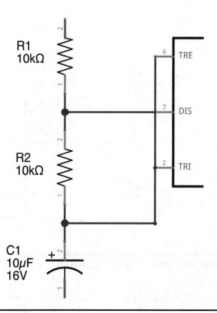

FIGURE 5-11   Connecting pins on IC1.

to the negative end of a battery for battery-powered equipment, and the other is the positive supply for the circuit. These nets usually connect to lots of parts in the design. As a schematic gets more complex, it can be tricky to connect everything with lines, especially these power nets.

To make things neater, special stand-in symbols are used for the GND and positive supply. You do not have to use them, but if you do, it makes for easier-to-read schematics. You will find these special symbols in the Schematic View section of the CORE parts bin. Drag on one ground connection and one V++ symbol. Change the voltage property of the V++ symbol to 3V, and connect them to the top of R1 and the bottom of C1, as shown in Figure 5-12.

Step 4. Connecting C2 and More Power Symbols

Drag a capacitor (ceramic this time) onto the schematic area to the right of and below the IC. Then duplicate the ground and 3V symbols (Duplicate is an option on the right-click menu, or type CTRL-D), positioning them to the right of IC1, then connect them. Duplicating a component or power symbol has the advantage that its properties are carried along with it, so you do not need to change the voltage to 3V for the V++ symbol.

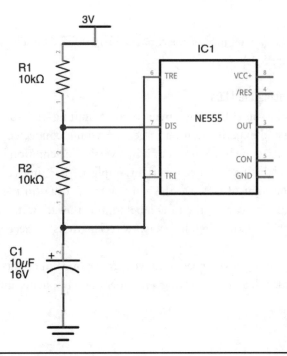

FIGURE 5-12   Adding supply symbols.

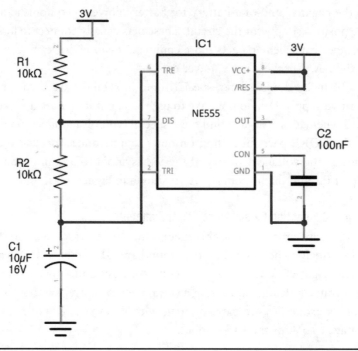

FIGURE 5-13 Connecting C2.

Set the capacitance of the capacitor to 100nF. The schematic should now look like Figure 5-13.

Step 5. Adding the LEDs

Connecting the LEDs and their current-limiting resistors may require a little patience and experimentation to get everything connected. First add one resistor (R3), set its value to be 68Ω, put it in a vertical orientation by rotating it, and then rotate its label back. Once it is right, duplicate it 3 times and spread the resistors out to the right of IC1. Do the same with an LED (in the CORE section of the parts bin). You can set the LED label to hide the color. It would be handy to have the color, but knowing its exact wavelength is unnecessary and clutters the diagram.

Incidentally, if you need to move a group of components, you can drag over an area of the schematic, selecting a number of components, and then drag the whole

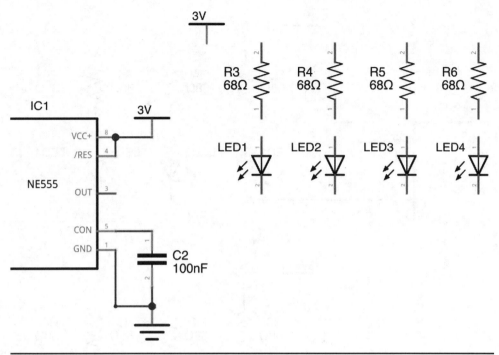

Figure 5-14   Lining up resistors and LEDs.

group. Similarly, you can duplicate a whole group of components by selecting them and then using the CTRL-D keyboard shortcut.

Line up the new components and a duplicate of the 3V supply symbol, as shown in Figure 5-14.

The way to connect a row of components to one net is summarized in Figure 5-15.

Start by connecting pin 3 of IC1 to the bottom of LED1 (Figure 5-15a). Next, drag from the bottom of LED2 back to the bottom of LED1 (Figure 5-15b). If you drag the other way around, then the net will disconnect from the bottom of LED1 and be moved to the bottom of LED2 rather than add a new line to the net. Repeat this process, connecting the bottom of LED3 to LED2 and finally LED4 to LED3 (Figure 5-15c).

Wire up the top of the LEDs to the 3V supply symbol, using the same approach, and join each LED to its resistor. The end result should look like Figure 5-16.

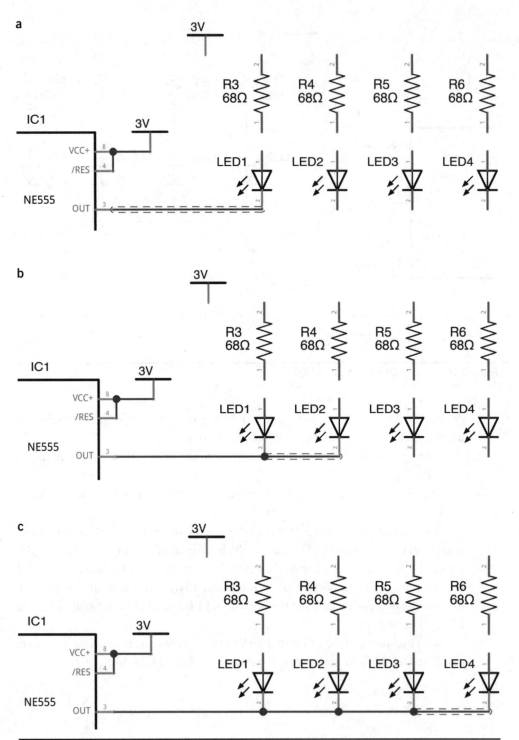

Figure 5-15 Connecting a row of components.

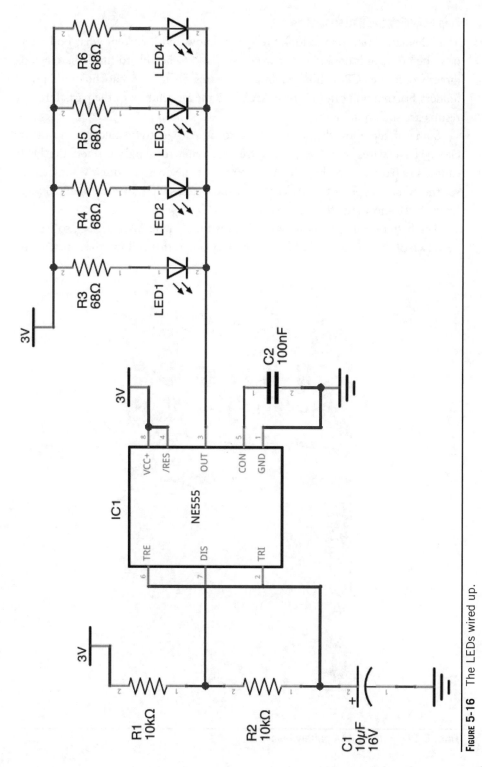

Figure 5-16 The LEDs wired up.

Step 6. Adding the Battery and Switch

The schematic is starting to look pretty good now. But it currently has no battery or switch to turn it on and off. It would be good to be able to fit the battery box directly onto the PCB without having extra wires. So, to see what choice of battery holders Fritzing will offer us, type **AAA** into the Search box of the parts bin. The results are shown in Figure 5-17.

Some of these are the wrong shape to be AAA battery holders, and others clearly have wires attached (which we are trying to avoid); so add the likely candidates from the results, and you can remove the ones you don't want later. To see the actual shape of the battery box candidates, switch over to the PCB view. Figure 5-18 shows the best of these.

The battery box shown in Figure 5-18 is for a single AAA battery, so we will need two of them. The symbol for this battery box rather follows the construction

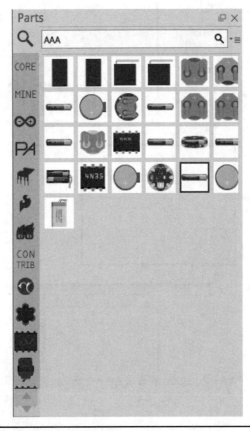

FIGURE 5-17   Finding AAA battery holders.

FIGURE 5-18   The battery box.

of the battery box itself that has two solder pins for each end of the battery. This is reflected in the symbol, which makes it less clear than the standard battery symbol and is reminiscent of an ancient Egyptian hieroglyph.

Switch back to the Schematic tab, and position the batteries over to the left of the schematic. Also add a switch from the Input section of the CORE parts bin. The switch is labeled Toggle Switch in the parts bin.

Finally connect everything and tidy up the labels, as shown in Figure 5-19. This is the same diagram that you saw in Figure 5-6. I have just repeated it here for convenience.

You will meet this design again in Chapter 6, where it is used as an example for PCB layout.

If you were designing this project as a real product, then you would find the battery box that you wanted to use first and either download (if available) a Fritzing part or create a custom part. You will learn how to make custom parts in Chapter 9.

Advanced Schematic Drawing

Although the bicycle light example covers most aspects of schematic drawing in Fritzing, there are a few techniques that it has not covered that are useful to know.

Net Labels

You have seen how you can use power symbols to reduce the number of lines on the schematic, keeping the spaghetti effect to a minimum. You can also use a similar approach with any nets on the schematic.

Rather than connect two (or more) points on a net with lines, you can add a label to both ends of the net to indicate that they are joined. Figure 5-20 shows a

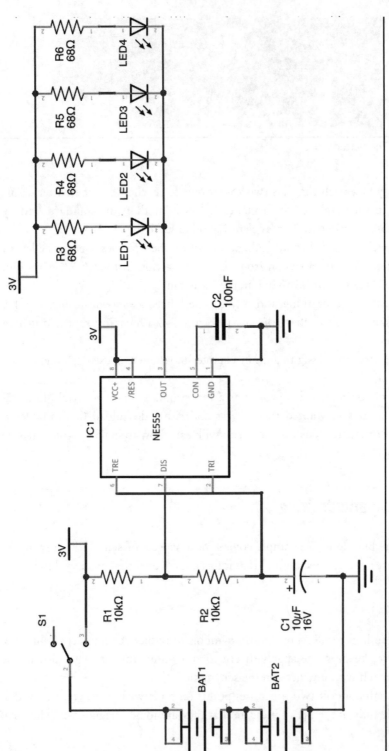

Figure 5-19 The complete schematic.

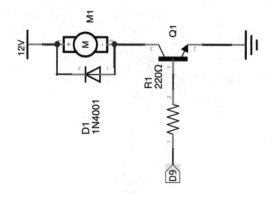

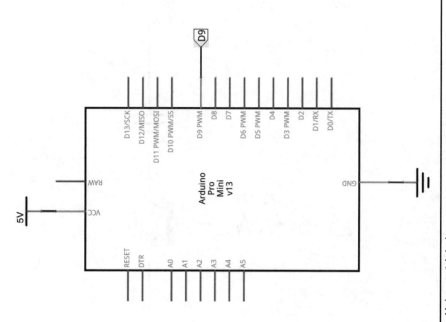

Figure 5-20 Using net labels.

design that uses this approach to connect an Arduino pin to an output circuit to control a motor.

The process of creating and linking a net label is summarized in Figure 5-21.

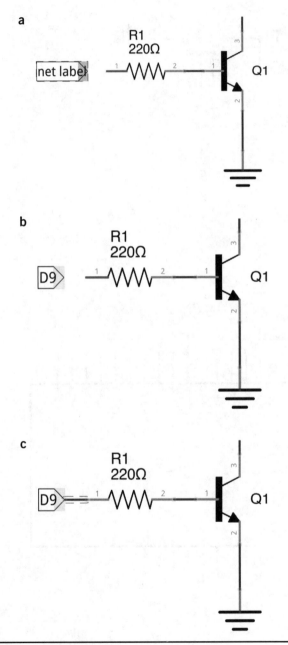

FIGURE 5-21   Creating a net label.

The first step is to add a net label from the Schematic section of the CORE parts bin (Figure 5-21a). Next, use the Inspector to change the label text. The label text does not have to match the net name (Figure 5-21b). Now connect the net label to the net or part terminal that you need to label, using the connection point at the pointy end of the net label (Figure 5-21c). If you connect the label before changing the name, then Fritzing will disconnect the net again when you change the label.

To create the other end of the link, simply duplicate the net label you just created, and then attach it to the other net. You can check that the two nets are actually connected by clicking on one of the component leads connected to one of the labels. For example, click on the left-hand connection to R1 in Figure 5-21. When you do this, both ends of the link will turn yellow for the duration of the click.

If other points in the schematic need connecting to the same net, then just repeat the procedure of duplicating the net label as many times as you need.

Frames

Electronic invention is an iterative process. Your design will get better as you refine the design through successive versions. Keeping track of exactly what version you are looking at will become important. You do not want to accidentally send off an old version of the design for fabrication. So it is important to name your designs carefully.

Including information about the design on the design itself is a good idea. Fritzing includes a special part called a *frame* for recording information about the design. You will find the frame part in the Schematic section of the CORE parts bin. Figure 5-22 shows the bike light schematic with a frame attached to it.

It's conventional to place the frame at the bottom right of the schematic. You can see from Figure 5-22 that the frame contains the Filename and Date that are completed automatically. The description in the topmost area and the project name are set by selecting the frame and then changing the "descr" and "project" attributes in the Inspector.

If you would prefer your frame box to actually have a border that will also enclose the whole schematic, then resize the frame part, dragging its top-left corner up and to the left to enclose the schematic.

The "rev" field is used to record the revision of the design. A common naming convention is to start with version 1.0 and a revision letter of a, b, c, etc. The letter changes are for minor internal changes that only you see. When something changes that other people will see—say, you are getting a new version of the PCB made—then bump the minor version up to 1.1. Changing the big number at the front is reserved for a completely new version of the product.

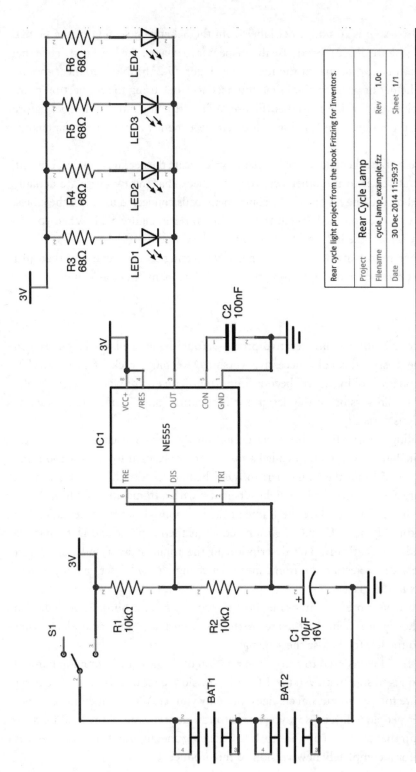

FIGURE 5-22 Adding a frame to a schematic.

Datasheets

Datasheets define exactly and reliably how a component behaves. These can be trusted to be correct. For simple components such as individual transistors, they tell you the safe limits for using the device such as the maximum current the transistor can switch before it gets too hot and that kind of thing. For more complex ICs, the datasheet becomes even more useful, as it will often include application notes that include example schematics for using the IC as well as guides for choosing the additional components that the IC may need. These designs are tried and tested, and it is a good idea to base your designs on these schematics.

As an example, take a look at the datasheet for the TLC555 IC. You can find this or any other component by typing the component's name followed by the word *datasheet* into your favorite search engine. Here is one link for the TLC555 datasheet from its manufacturer, Texas Instruments: www.ti.com/lit/ds/symlink/tlc555-q1.pdf.

Often you can find out what you need from the summary information at the top of the datasheet. Figure 5-23 shows this for the TCL555.

On the right-hand side you can see that the operating voltage is 2 to 15V, which is good. Also, on the left-hand side, you can see the words *High Output-*

TEXAS INSTRUMENTS   **TLC555-Q1**

www.ti.com   SLFS078A–OCTOBER 2006–REVISED OCTOBER 2012

LinCMOS™ TIMER
Check for Samples: TLC555-Q1

FEATURES

- **Qualified for Automotive Applications**
- **Very Low Power Consumption**
  - **1 mW Typ at V_{DD} = 5 V**
- **Capable of Operation in Astable Mode**
- **CMOS Output Capable of Swinging Rail to Rail**
- **High Output-Current Capability**
  - **Sink 100 mA Typ**
  - **Source 10 mA Typ**
- **Output Fully Compatible With CMOS, TTL, and MOS**
- **Low Supply Current Reduces Spikes During Output Transitions**

- **Single-Supply Operation From 2 V to 15 V**
- **Functionally Interchangeable With the NE555; Has Same Pinout**

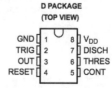

D PACKAGE
(TOP VIEW)

GND	1	8 V_{DD}
TRIG	2	7 DISCH
OUT	3	6 THRES
RESET	4	5 CONT

FIGURE 5-23   Datasheet for the LTC555.

Current Capability under this, as the datasheet claims a sink current of 100mA. This is more than enough for our four LEDs.

Farther down the datasheet, you will also find a section "Absolute Maximum Ratings." This tells us that actually, we could get away with sinking 150mA if we really needed to. It is worth reading through the whole datasheet even if you do not understand everything on the datasheet. It will give you an understanding of just what the chip is capable of doing.

Summary

In this chapter, you have learned both how to neaten up a datasheet generated by Fritzing from a breadboard design and how to create a schematic from scratch.

In Chapter 6 you will learn how to create a PCB design for the rear bicycle light project.

PCB Layout

In this chapter, you will learn how to design a PCB by using Fritzing. The cycle light project used as an example in Chapter 5 is continued and developed into a full PCB design. Both through-hole and surface-mounted versions of this design are developed.

Printed Circuit Boards

PCBs come with their own set of jargon, and it is worth establishing exactly what we mean by *vias, tracks, traces, pads,* and *layers*.

The main focus of the book will be on making double-sided professional-quality boards, either by using the Fritzing integrated PCB service or by e-mailing the design files to a low-cost PCB fabrication service (as low as $10 for 10 boards). The making of PCBs at home is now hardly worthwhile because fabrication services can make PCBs at lower cost and to a better standard than home PCB etching, with all its attendant problems of handling and disposing of toxic chemicals or need for expensive milling machines.

Figure 6-1 shows the anatomy of a two-layer PCB.

Refer to Figure 6-1 for the following discussion.

- *Pads* are where the components are soldered to the PCB. So for through-hole components, each pad will have a hole in the center for the component lead to poke through.
- *Tracks* or *traces* are the copper tracks that connect the pads. There can be tracks on both the top and bottom surfaces of the PCB. The traces are

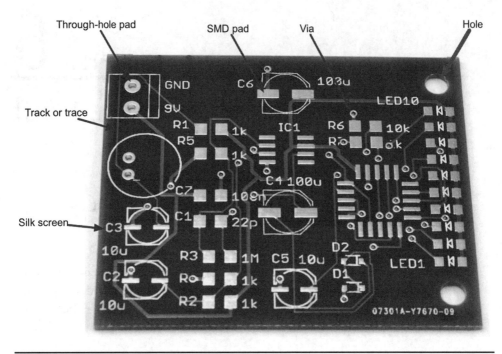

FIGURE 6-1   The anatomy of a double-sided PCB.

insulated by a layer of lacquer called *solder mask*, over the entire surface of the PCB, except where there are pads to be soldered.

- *Vias* are small holes through the board that link a bottom and top trace electrically. Traces on the same layer cannot cross, so often when you are laying out a PCB, you need a signal to jump from one layer to another.
- *Silk screen* refers to any lettering that will appear on the final board. It is common to label components and the outline of where they fit, along with the component's name, so that when it is time to solder the board, it is easy to see where everything fits.

Cycle Light Example

The main example project in this chapter is the rear cycle light project that was developed in Chapter 5. So if you were following along with that design, open it up again now. You can also load the finished design into Fritzing as the file is called cycle_lamp_example.fzz and is included in the downloads for the book that are available from the author's website at www.simormonk.org.

Step 1. Spread Out the Parts

When you first switch over to the PCB tab on your design, you will be greeted with something like Figure 6-2.

It even looks as if some of the LEDs and resistors are missing. Actually they are there, but they are just stacked on top of one another. So the first task is to spread out all the components so you can see what's going on. Figure 6-3 shows the parts spread out, but in no particular order.

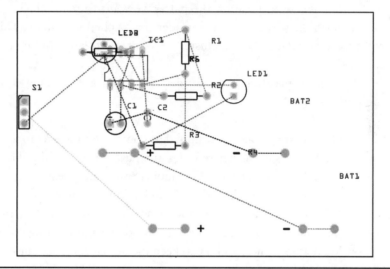

FIGURE 6-2   The initial PCB layout mess.

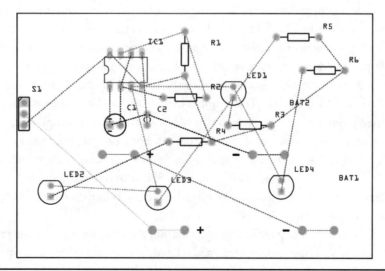

FIGURE 6-3   The components separated on the PCB.

When trying to drag the components apart, you will probably find that you cannot help but hit the dotted lines on the diagram, causing them to turn into orange-colored PCB traces. It's much too soon to be laying out traces, so when this happens, use the Undo button or CTRL-Z (ALT-Z on a Mac) to undo the trace creation. To help with this problem, you can hover the mouse over the area where you want to click, and a rectangle will appear showing what you are about to click.

Step 2. Place the Parts in Position

In deciding where to position the parts, we want to make their connection as easy as possible, so we need to consider the original schematic. This means we will probably want to have a cluster of components surrounding the integrated circuit (IC).

These are some things to consider when you are deciding where to place components:

- Note the physical constraints that come from the eventual enclosure for the project such as where LEDs and pushbuttons need to be placed.
- Consider the functional grouping of components that are heavily interconnected, such as the IC and its timing components.
- Minimize the messiness of the rat's nest, and try to reduce the number of times lines cross one another by positioning and rotating the components.
- Make it look nice, and try to keep the design compact but neat.

The LEDs need to be in a row across the middle of the PCB, spanning the whole width, and it makes sense for each of the current-limiting resistors to be near their respective LEDs. Figure 6-4 shows an initial positioning of the components. The components have also been rotated where necessary and labels moved around in just the same way as you did for schematic symbols in Chapter 5.

By convention, it is nice to have the IC with the notch at the top of the PCB, so that pin 1 is top left. This is not essential, but can help when you are first getting used to laying out a PCB. Also you will sometimes see a component that would be much easier to connect if it were rotated. For example, the initial orientations of C1 and IC1 are shown in Figure 6-5a. As you can see, the trace lines are crossing. By rotating C1 through 180° (Figure 6-5b), converting the dashed lines into PCB traces becomes trivial. Incidentally, these dashed lines *are* referred to as the *rat's nest* or sometimes *air wires*.

When you are positioning and rotating the components, use the rat's nest lines to show you how components are related. Try to minimize the crossing over and long length of the lines.

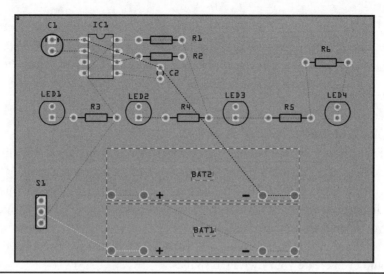

Figure 6-4   Initial component positions.

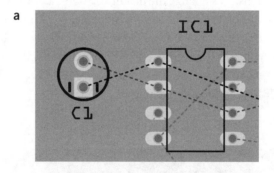

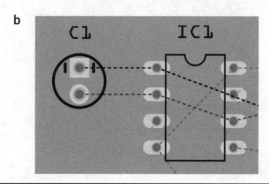

Figure 6-5   Rotating components for easier connections.

TIP *To stop the PCB as a whole from moving around if you accidentally click on it, you can lock it in position by selecting the gray outline of the PCB away from any components. Right-click and select Lock Part from the menu that pops up. You can always unlock it later if you need to move it.*

Step 3. Resize the PCB

Even though in this case the default PCB size is quite good for this design, you could make it a little smaller. The cost of PCB fabrication is usually based on the surface area of the PCB. Also, some PCB fabrication companies have standard dimensions where, at least at the prototyping stage, the PCBs are much cheaper if you limit them to a certain size.

Later in this chapter, you will learn how to use other predefined PCB shapes and even create your own PCB shapes.

Select the PCB as a whole by clicking on the gray background away from any parts, and the Inspector will show the width and height of the PCB. In this case, the width is 84.7mm and the height is 56.4mm. Since 50 and 100mm are common cutoff dimensions, let's change the height to 50mm and the width to 100mm. That way the battery holders can be side by side, making for a slimmer design.

Select groups of components and shuffle everything to fit within the new PCB border. The switch has moved over to the top right, and the batteries are rotated through 180° so that the positive terminals are to the right. When you have done that, the PCB should look something like Figure 6-6.

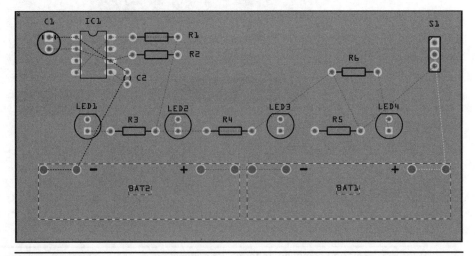

FIGURE 6-6 All the components are in place.

Step 4. Add Fixing Holes

The PCB will eventually need to fit into a plastic enclosure, and this will be easier if we put some holes in the PCB so that we can attach it to the plastic case with screws. Holes are added to the PCB as if they were components, but obviously they have no electrical connections. You will find the hole part in the PCB View section of the CORE parts bin. This is not a big PCB, so two holes should be enough to anchor it securely onto whatever enclosure is used. Drag two holes onto the PCB, either side of the LEDs.

Use the Inspector to change the hole diameter to 3mm. Another property of the hole that you may want to change in the Inspector is the ring thickness. This will provide a copper ring and area clear of copper coated with solder around the hole. This can be used to make electrical connections to a screw through the hole, but in this case there is no need other than I think it looks good.

If this were a real product, you would probably select the screws that you want to use before selecting the hole size.

With the holes added (Figure 6-7), the PCB is now ready for us to start laying down some copper.

Step 5. Route the Bottom Layer

Fritzing has an Autorouter, which will lay out the traces on your board for you. Try it, if you like, by clicking on the Autoroute button at the bottom of the Fritzing window. If you don't like the results (and you probably won't), just click Undo. Personally, I think it is worth taking the time and effort to route a PCB by hand.

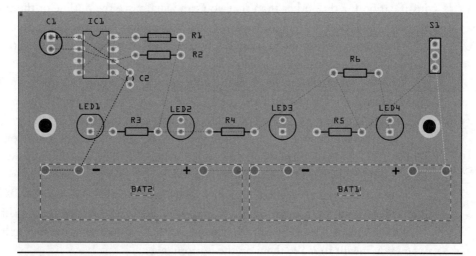

Figure 6-7   The PCB with mounting holes.

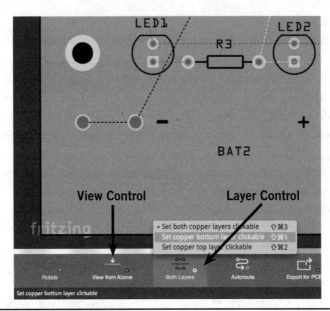

FIGURE 6-8   The view and layer controls.

It acts as an additional check to ensure the design is right as you work through each connection that needs to be made. It's also fun; it's like completing a puzzle, getting all the connections made without any of the traces on the same layer crossing over one another.

When laying out a through-hole design like this, I prefer to put much of the wiring into traces on the bottom layer of the PCB. I do not really have a good justification for this, other than that you can usually see the outline of the traces through the lacquer layer and so it just makes the top layer a tiny bit neater.

To control the routing, you use the two controls at the bottom of the Fritzing screen highlighted in Figure 6-8.

With lots of traces on the PCB, it can be difficult to see what's going on. The View control allows you to view the PCB from above or below. Click on it to toggle between the two. Most of the time, you will want to leave this as View from Top. Even when viewing from the top, you will be able to see the traces on the underside of the board. Traces on the top of the board will be shown in yellow and those on the bottom of the board in orange.

The Layer control constrains which layer traces will be drawn on. When you are working on the bottom layer traces, set it to Set Copper Bottom Layer Clickable. This will prevent you from accidentally creating traces on the top layer when you meant to make them on the bottom layer.

Select Set Copper Bottom Layer Clickable and start making the really obvious connections. Remember, the goal is to change all the rat's nest lines (air wires) into solid copper traces. Use the zoom feature if you are struggling to see where connections are going in a crowded area of the PCB.

Looking at the PCB area around R4 (Figure 6-9a), you can see that there appears to be an air wire connecting both pins of R4. This can't be right, it just looks that way. Because the air wires have no bend points, they just go straight from one pin to the next. To reveal where the connections really need to go, temporarily move LED2 down a little (Figure 6-9b). You can see that the right-hand pin of R4 is connected to the right-hand pin of R3 (just to the left of LED2), and the line just happens to run through the left-hand pin of R4 because we lined up the components nicely. We will come back to these trickier connections later.

Now that we are starting to see more complex lines, I changed the background color for the PCB view to be white. Your Fritzing will have a gray background, but you can change this from the View menu if you like.

Set the View control to View from Above and the Layer control to Set Copper Bottom Layer Clickable, and make all the obvious straight-line connections on the bottom surface of the board. Double-clicking on an air wire will turn it into a trace. You can also drag out from one end of the air wire.

You may find that some connections look as if they should route in a straight line, but then Fritzing inserts a bend point for you. This happens around the battery holder pins, as shown in Figure 6-10. If this happens, delete the trace (or CTRL-Z it) and drag it out, again holding down the SHIFT key. You can also remove a bend point by double-clicking on it.

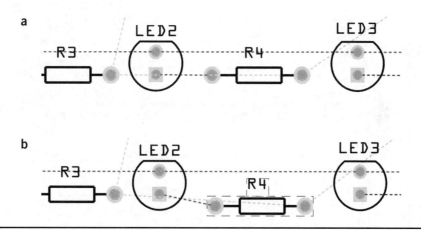

FIGURE 6-9 Untangling air wires.

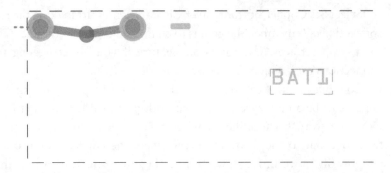

Figure 6-10 Unnecessary bend points.

Figure 6-11 shows the PCB with the obvious straight lines all routed. If you make a mistake, then CTRL-Z will undo the last action. You can also select a trace and then press the DELETE key or select Delete from the right-click menu. This will delete just the trace, not the underlying need for a connection. In other words, this will not affect the schematic. Do not be tempted to use the Delete Rat's Nest option as this will change the circuit. If you do this and flip over to the Schematic view, you will find that a line has vanished.

So far, so good. Staying with the bottom layer, make the trace between pin 3 of IC1 and the top connection of LED1. This needs to come out and away from pin 4 of IC1. Let's keep the traces on the PCB more or less vertical or horizontal for the sake of neatness.

Start at the IC1 pin 3 end of the connection because this does not have other connections branching off it as the LED1 end does. Drag from the air wire near

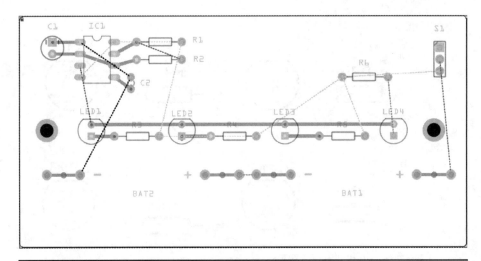

Figure 6-11 The "low-hanging fruit" traces on the bottom layer.

the pin rather than from the pin itself. Start to drag out to the left from IC1 pin 3, and a bend will appear in the air wire (Figure 6-12a). This is great, because now we can see exactly the intended destination of the trace. Release the mouse button and a trace will be drawn (Figure 6-12b). This is okay, but let's neaten it up by adding another bend point and then moving both bend points. Figure 6-12c shows the final result of making this trace.

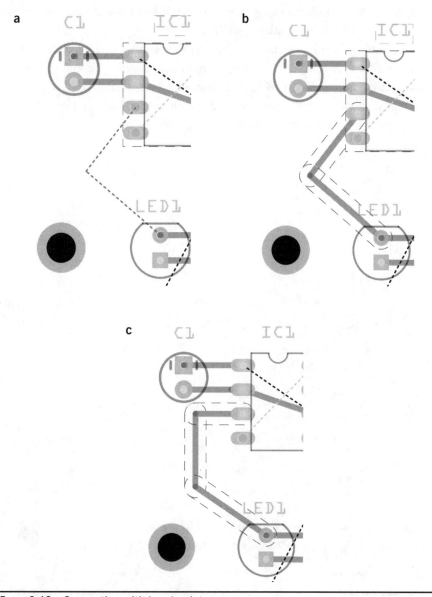

FIGURE 6-12   Connecting with bend points.

I have gone with an angled trace segment for the last bit of the connection to LED1 to keep the trace well away from the hole to the left of LED1.

Another tricky routing problem arises when you want to route the traces one way, but the rat's nest air wires are steering you in another direction. The air wires always take the shortest-distance route from one connection to another. For example, in Figure 6-13a, two air wires go from the right-hand terminal of R4 to the left-hand terminal of R6 and then the right-hand terminal of R5. In this case, it would be a whole lot easier to route from R4 to R5 and then connect the right-hand pin of R6 to the bottom pin of the switch.

Making these connections in the way that you want can take a bit of trial and error. One way to force this route, temporarily drag R6 up and to the right. This will cause the rat's nest to rearrange itself as we would like (see Figure 6-13b). Finally, we can wire the air wires and then move R6 back and rearrange the bend points to get the result shown in Figure 6-13c.

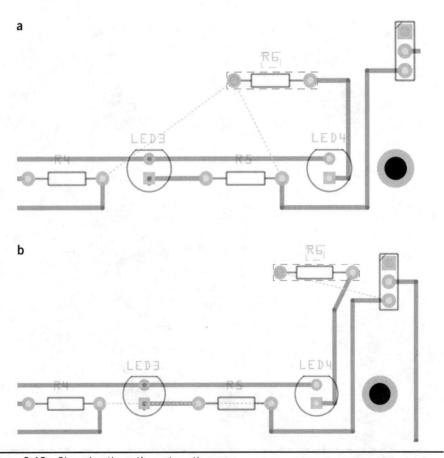

Figure 6-13   Changing the rat's nest routing.

c

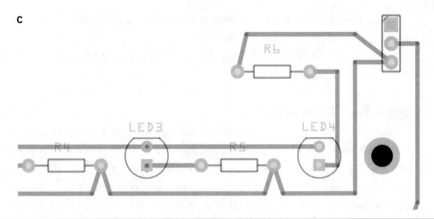

FIGURE 6-13 Changing the rat's nest routing (*continued*).

You can also route away from air wires, if you click on one of the alternative connection points that will be highlighted in yellow as you start to drag out an air wire.

Repeat this process for all the other connections that can be made on the bottom surface of the PCB, without crossing any of the existing traces. Use Figure 6-14 as a guide. Note that when you come to route the right-hand connection of R1, Fritzing is suggesting that we route this to the right-hand connection of R3. We can't do this without crossing a trace, so it would be much easier to connect it to the left-hand pin of R6. To make the air wire reroute, temporarily drag R6 over toward R1, make the trace, and then drag R6 back to its original position.

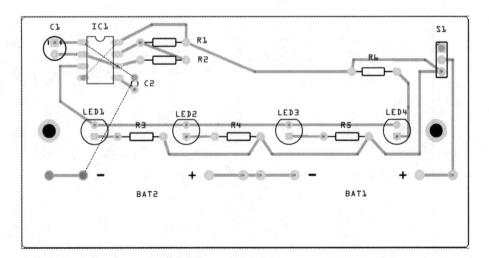

FIGURE 6-14 All the bottom traces routed.

There is no need to worry about traces crossing over the component labels, as these will be printed on top of the traces.

If you have arrived at the stage shown in Figure 6-14, there are only three air wires left to be routed.

Step 6. Route the Top Connection

Looking carefully at the connections that remain, you can see that the top pin of C2, pin 1 of IC1, and the negative battery connection are all ground connections. We will handle these, using something called a *ground fill*, in the next step. This just leaves one connection to be made on the top surface of the board, between pins 4 and 8 of IC1. To make this connection, select the option Set Copper Top Layer Clickable on the Layer control, and then join the pins on IC1 as shown in Figure 6-15.

Note that the trace on the top layer is shown in a mustard yellow color rather than the more orange color of the bottom layer traces. It does not matter that the trace between pins 2 and 6 of IC1 crosses the trace between pins 4 and 8 because one is on the top layer and the other is on the bottom.

Step 7. Add a Ground Fill

It is a common practice in PCB design to use a ground fill (sometimes called a *ground plane*). This ground fill is optional and can be on one or both of the copper layers. A ground fill makes all the layer one big area of copper connected to the ground net, apart from channels in the ground fill for traces that are on that layer, but not connected to ground.

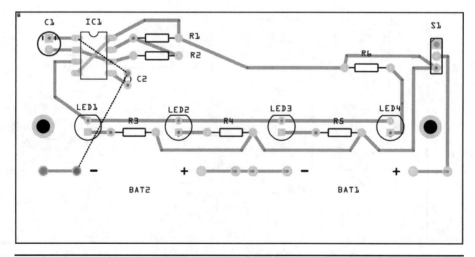

Figure 6-15   Routing the top layer.

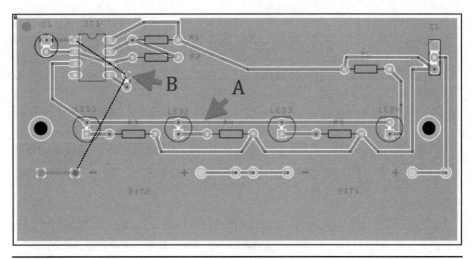

Figure 6-16   Adding a ground fill to the bottom layer.

You will use this ground fill as a way of connecting the three remaining unrouted pins.

Set the View control to be View from Above and the Layer tool to Bottom Layer. Now from the Routing menu select the options Ground Fill and Ground Fill Bottom. The board should now look like Figure 6-16.

There are a couple of things to note in Figure 6-16. First, the non-ground traces are now in a channel, with a copper-free narrow gap on each side of the traces (A). Second, if you look closely at the negative battery connection, the top pin of C2, and pin 1 of IC1, you can see that they all have little traces from the ground fill to the pad (marked with B). Figure 6-17 shows a close-up of this for the battery terminal.

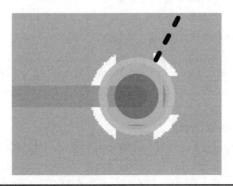

Figure 6-17   A pad connected to the ground fill.

There is an interesting reason why the ground fill only connects to the pad on each side, rather than simply covering the pad. The reason is that when you solder a pad, it has to get hot; and if the pad is effectively the whole ground plane, then it is much harder to get the part of the pad hot that needs to be hot, as a lot of heat is carried away into the surrounding copper.

The other thing that you will notice from Figure 6-16 is that the air wires for the ground connections are still there despite the connections having been made by the ground fill. This is a "feature" of the version of Fritzing used here (Version 0.8.7), and this may well have been fixed in your version. There are two work-arounds for this:

- If you are sure that the connections are actually made by the ground-fill layer, then you can simply hide those few remaining air wires from the View menu by deselecting the Rat's nest Layer.
- If you explicitly connect up the remaining air wires as traces, before adding the ground fill, this will also remove the air wires.

At this point, it is a good idea to look carefully at the ground fill and make sure that there is some route between the pins relying on it. The only pin you might have doubts about is the top of C2, as the ground fill here looks a bit like an island. However, there is a connection, because the ground fill copper seeps between the pins of IC2, so all parts of the ground fill are connected.

Step 8. Add Text and a Logo

The PCB is starting to look pretty professional. As a final touch, add some text to identify the PCB version and for good measure a logo graphic.

To add text, just drag a Text part from the PCB View section of the CORE parts bin onto the PCB. You can then edit the text in the Inspector. If you want to add an icon too, then drag a Silkscreen Image from the parts bin. There are a few different icons to select from in the Inspector. I have chosen the Open Hardware logo, to show that this design is Open Source. The end result of this is shown in Figure 6-18.

The rat's nest has vanished because, having assured myself that they have actually been routed, I have deselected Rat's Nest Layer from the View menu.

Step 9. Design Rule Check

On a simple design like this, it is fairly easy to manually check the design for traces crossing on the same layer, running too close to pads, etc. However, it is always worth running the Design Rule Checker (DRC), which will automatically check

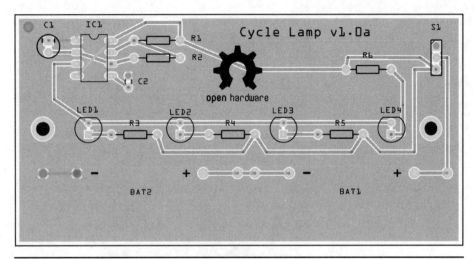

Figure 6-18   The PCB design with text and logo.

for these things. You can run this from the Routing menu. Once it has run, if all is well, you will get a message that says: "Your sketch is ready for production: there are no connectors or traces that overlap or are too close together."

Hurray, your PCB layout is good to go. In Chapter 7, you will learn how to go about having real PCBs made from your design as well as how to solder components to them. But before that, there are quite a lot of details that we skipped over in this example that I want to catch up on.

Advanced PCB Layout

In using the cycle lamp PCB example, there are a few features of Fritzing that we haven't needed to use and some aspects that we have only touched on. In this section we will fill those gaps.

Moving Traces Between Layers

If you draw a trace on the bottom layer and then decide it would be better on the top, you can right-click on the trace and select the option Move to Top Layer. A similar option will appear in the menu when you have a trace on the top layer that you want to move to the bottom.

You can also change the layer of a trace by selecting it and then using the Inspector (Figure 6-19). The Inspector also allows a number of other things about the trace to be changed.

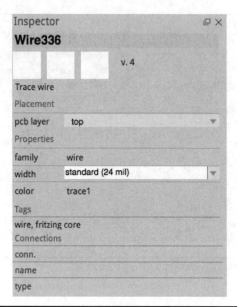

<figure>**FIGURE 6-19**   Selecting a trace layer from the Inspector.</figure>

Changing Trace Widths

The wider a trace, the more current it can carry without overheating. By default, Fritzing uses a trace width of 24 mils. The unit *mil* is often confused with mm. These are not the same. One mil is 1/1000 inches. In modern PCB design, 24 mils is actually a medium width. Many PCB manufacturers will allow you to use traces down to 6 mils. If there is plenty of room on the PCB, then 24 mils is a good width to use for most purposes.

In a later section of this chapter, you will see how for surface-mounted components, 24 mils can be a bit too wide, and you will need to use something thinner.

Where you have a higher current part of the PCB design, you should use thicker traces.

Table 6-1 provides a rough guide to the current-carrying capabilities of various trace widths. These figures are based on a maximum allowable increase in temperature of the trace of 10°C at an ambient temperature of 25°C and a 100mm trace length.

The data for Table 6-1 are derived from the useful online calculator at http://circuitcalculator.com/wordpress/2006/01/31/pcb-trace-width-calculator/.

TABLE 6-1   Current-Carrying Capabilities of Various Track Widths

Trace Width	Copper Layer Thickness	
	1oz/in² (0.035mm)	2oz/in² (0.07mm)
8 mils	700mA	1.2A
12 mils	1A	1.7A
16 mils	1.25A	2.1A
24 mils	1.7A	2.8A
32 mils	2A	3.4A
48 mils	2.7A	4.6A

For historical reasons, the most common thickness of copper on PCBs is 1oz/ft². That is 1oz of copper spread over 1ft². In other words, it is pretty thin (about 0.035mm). PCB manufacturers will often offer a thickness of 2oz/ft² (0.07mm). Just stick to 1oz/ft² unless you have a good reason to do otherwise.

As you can see from Table 6-1, even a very thin 8mil trace can carry a fairly sizable current before getting warm enough for any long-term damage to the PCB to occur.

Placing Components on the Bottom Layer

At present, all the components for the cycle lamp are mounted on the top surface of the PCB. The board could be made a little smaller if the battery holders were mounted on the underside of the board. The obvious advantage of putting components on both sides of the board is that it saves space and allows you to make a smaller PCB. The disadvantage is that it may increase manufacturing costs.

Switching a component itself from the top layer to the bottom layer is quite straightforward. You select the components (for example, BAT2) and then in the Inspector select Bottom from the PCB Layer drop-down list. Do this for both BAT1 and BAT2, and the resulting layout will look like Figure 6-20a. Notice how when you look from above, the labels for BAT1 and BAT2 are in mirror image. It is as if we are looking through the top surface of the board.

You can see in Figure 6-20a that as you might expect, this has messed up the routing of traces considerably. So remove the ground fill and delete all the wires around the battery holders so that we can see what's going on (Figure 6-20b).

In Figure 6-20b, most of the layout mess stems from the fact that now the battery holders are on the other side of the board—their positive terminals are

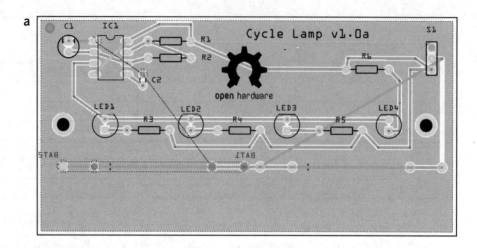

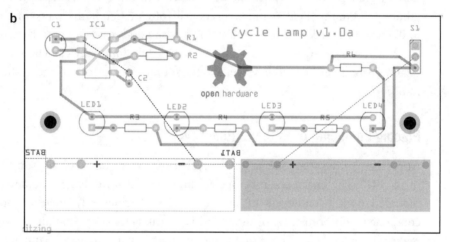

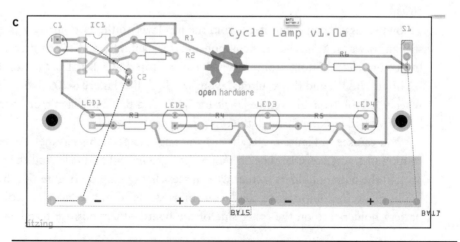

FIGURE 6-20 Moving the battery holders to the underside of the PCB.

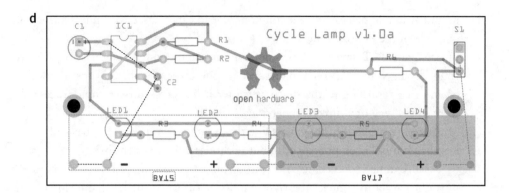

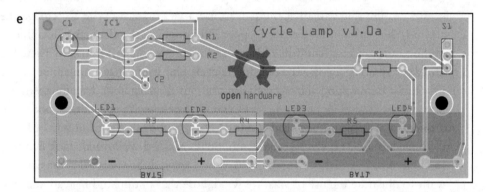

Figure 6-20   Moving the battery holders to the underside of the PCB (*continued*).

now to the left of the board rather than to the right. In moving the battery holders to the bottom layer, they have been flipped from left to right. To fix this, we can rotate both battery holders by 180° (Figure 6-20c). Thus the body of the battery holders is toward the center of the PCB, allowing us to make the PCB a bit smaller, as there is now an empty area at the bottom.

Now move the mounting holes and the battery boxes up a bit and resize the PCB (Figure 6-20d). In moving the left-hand mounting hole, the trace from pin 3 of IC1 also had to be moved a little. The labels for BAT1 and BAT2 need to be moved as well.

Route the remaining air wires and reapply the ground fill. The final PCB is shown in Figure 6-20e.

Vias

In the cycle lamp example, there are not so many parts, and they are well spaced out, so routing is pretty easy. Also since it is a though-hole design, there are naturally lots of places where a component lead can be used to connect a trace on one side of the board to the other.

In many PCB layouts, especially on surface-mounted designs, you need to route an air wire as a trace that starts on one layer and then swaps to another layer, often returning to the first layer after it has crossed over some traces that were in the way of a direct connection.

To connect between the layers like this, you need to use a via. Basically a via is a through-hole component lead, but without the component itself.

As an example, let's change the routing on the cycle lamp project to deliberately introduce a via to connect the right-hand side of R5 to the left-hand side of R6, without having to visit S1 on the way (Figure 6-21a shows the old and new routes).

To reroute this part of the design to use a via, first you need to remove the ground fill by using the Remove Copper Fill option on the Ground Fill submenu of the Routing menu. You will add the ground fill back again later. Also delete the traces that you are going to change by right-clicking on the old route and selecting the Delete Wire option. The result of this is shown in Figure 6-21b. Interestingly, the air wire now shows a direct connection in the route that we want to take. You cannot simply drag a trace on the bottom layer from the right-hand lead of R5 to the left-hand lead of R6, because that would cross the horizontal track between LED3 and LED4. In this case we could route a track on the top layer, but since we want to illustrate the use of vias, we will ignore that option and instead add two vias from the PCB section of the CORE parts bin (Figure 6-21c). Position one via between the right-hand lead of R5 and the bottom lead of LED4. Add the second via vertically above the first via, but on the other side of the horizontal line that is in the way. We are going to use the pair of vias as a kind of bridge over the horizontal track.

Set the Layer control to Bottom Layer and drag a trace from the bottom via to the right-hand connection of R5. Then drag a trace from the top via to the left-hand connection of R6 (Figure 6-21d). Finally, flip the Layer control to the top layer, and drag a trace on the top layer from one via to the other. The final result is shown in Figure 6-21e. You can then reapply the ground fill.

You will meet vias again when we make a surface-mounted version of the cycle lamp.

Icons Revisited

Fritzing is supplied with a few ready-made icons, such as the OSH logo, which you can add to your PCBs. You can also add your own custom logo to the PCB. The first step in adding your own logo is to create an image file.

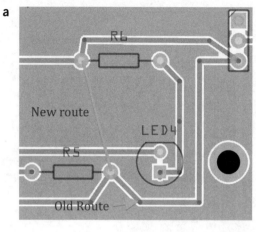

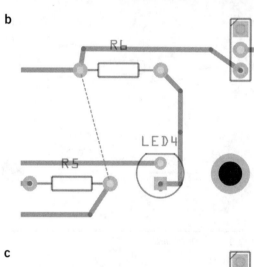

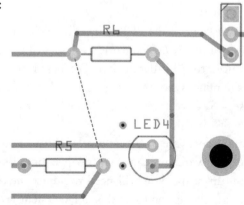

FIGURE 6-21   Using a via.

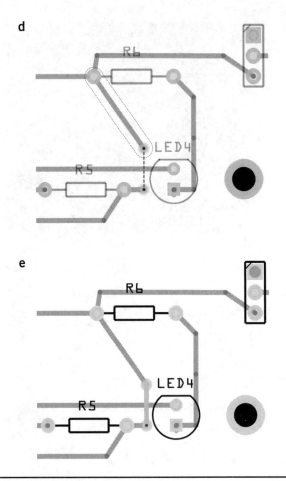

FIGURE 6-21   Using a via (*continued*).

The silk screen layer of the PCB is usually made up of white onto the green solder mask layer of the PCB. There are no shades of gray. So any image that you add will be reduced to a bit depth of 1. For best results you should design the image with this in mind. Once the image is added, you will be able to resize it. For best results, use a SVG vector image, although Fritzing will also accept a PGN format image.

There are many software packages available for creating images like this, and the open-source software Inkscape will do just fine. Once you have created a suitable image in Inkscape, you will need to adjust the document size to fit the image. You will find the option to do this in Document Properties in Inkscape (see Figure 6-22). Now, add an Image to the PCB and click on the Load Image File button to select the image file. Resize and move it, and your icon will be on the board, as shown in Figure 6-23.

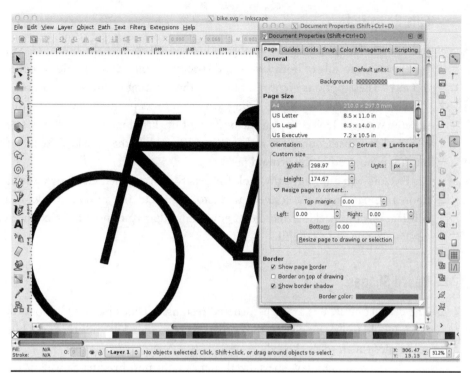

FIGURE 6-22   Adjusting the document size in Inkscape.

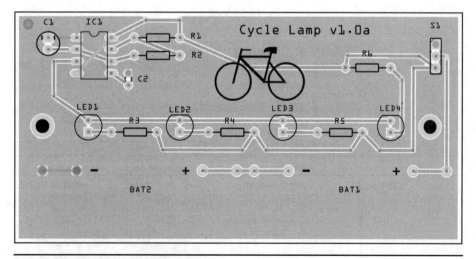

FIGURE 6-23   A custom icon on the PCB.

Fills Revisited

In the cycle lamp project, we applied a ground fill to only the bottom layer. You can, in fact, add fills to both layers and specify whether they should be connected to GND. These options are available from the Routing menu (Figure 6-24).

They layers available for filling will change depending on what you have selected with the Layers tool at the bottom of the Fritzing window. If you select an option that is just Copper Fill rather than Ground Fill, then the layer will be filled, but not connected to the ground net.

Fritzing will automatically look for nets that are named GND or Ground and upon finding them will connect them to a ground fill automatically. You can confirm this by checking that a component pin that you know to be connected to GND is connected to the fill.

Board Shapes

The cycle lamp PCB is a sharp-cornered rectangle. That's just fine, it does the job, but it would look a bit nicer if the corners were rounded. It's a little thing, but does add a nice touch to the board.

There are quite a few specialized shapes built into Fritzing as well as options for Rectangle and Round Rect. You can access these by selecting the board itself, by clicking well away from any parts; then in the Selector, click on the Shape drop-down list. Select Round Rect and the cycle light board will get rounded edges. The ground fill will, however, still be rectangular, so remove the ground fill and then add it again. Now the board should look like Figure 6-25.

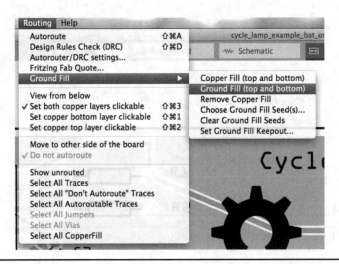

FIGURE 6-24   Fill options.

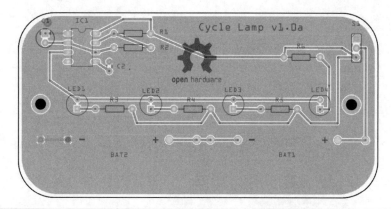

FIGURE 6-25   Adding rounded edges to the PCB.

Fritzing also allows you to create boards of any shape you like by supplying it with an SVG image file. This image is a bit special in that it must have two layers. One layer is for the board outline itself, and one is for the available silkscreen layer. Both layers will normally be of the same shape and on top of each other. The shape on the board layer needs a solid fill color (by convention green) and no stroke. The silkscreen layer must have no fill and a white stroke of 0.2mm, which makes it a little tricky to see.

To create such a shape in Inkscape, perform the following steps:

1.   Create a layer and name it "board."
2.   Draw your shape at the size that you want on the board level.
3.   Set the shape's fill color to green, and set its outline to have no stroke.
4.   In the Layers area (bottom right of Figure 6-26), select the board layer; then right-click on it and select the option Duplicate Current Layer.
5.   Rename the current layer *silkscreen.*
6.   Select the duplicated shape on the silkscreen layer, and set its stroke to 0.2mm white. Next, set its fill to None.
7.   Save the file. Select the Save As... menu option, and in the file dialog select the file type of Plain SVG (*.svg) before saving the file.
8.   Fix the saved file in a text editor.

Step 8 is necessary, because unfortunately when Inkscape saves a layer, it does not transfer the name attribute that you specified for the layer (that is, "board" or "silkscreen"). So open the image file in a text editor, and search for the first "g" tag, something like the following example:

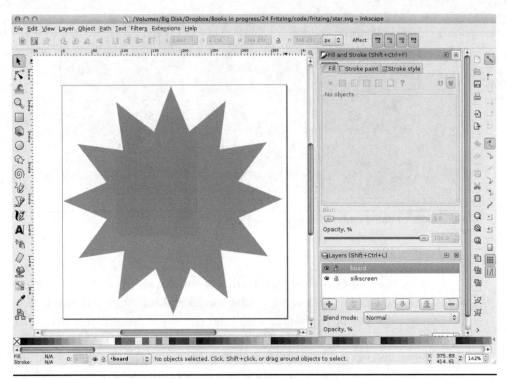

Figure 6-26 Creating a custom PCB shape in Inkscape.

```
<g
    id="g3860"
    style="display:inline">
    <path
```

Replace the value of the id attribute to be "silkscreen," as shown below:

```
<g
    id="silkscreen"
    style="display:inline">
    <path
```

A little farther down the file there should be another "g" tag. Change the id of this tag to "board" and then save the file.

Now that you have a file in the correct format, you can apply that shape to any PCB, by selecting the PCB in the PCB tab and then clicking on the Load Image File option. You will notice that a dialog pops up, asking you to check the Gerber production files before you rely on this feature working. Checking the Gerber files in a Gerber viewer is always a good idea until you become really confident with

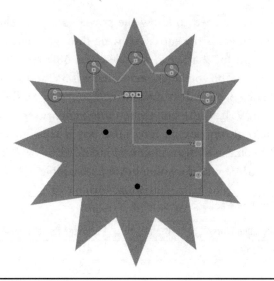

FIGURE 6-27   A star-shaped PCB.

Fritzing; this is discussed in Chapter 7. An example PCB using a starting shape is shown in Figure 6-27. You can find both the image file (star.svg) and the Fritzing file (ch_06_PCB_Shape.fzz) in the downloads available for this book.

SMD PCBs

Surface-mount devices (SMDs) cost less and are smaller than the through-hole components we used to develop the cycle light project. We could actually take the design and modify it to use mostly surface-mount components.

SMD Components

Not all the components we used in the project have surface-mount counterparts. Take, for example, the battery holders. Also some of the components might be better as through-hole. The LEDs will have bigger, better lenses as 5mm through-hole than tiny surface-mount devices. However, for the sake of illustrating a surface-mount layout, let's replace everything we can with SMDs.

Resistors

Starting with the resistors, we have used ¼W resistors. This is an indication of the power the resistor can handle, and ¼W is fine. Before choosing SMD equivalents of the resistors, we need to know the power rating we should aim for with the resistors. The power is the current flowing through the resistor, multiplied by the voltage across it. In Chapter 5, we selected a 68Ω resistor that would give us a current of 20mA. The voltage of the batteries is 3V, and 1.7V of that will be across the LED, leaving 3 − 1.8 = 1.3V across the resistor. So the power will be 1.3V × 20mA = 26mW. If you want to know the theory, then take a look at http://en.wikipedia.org/wiki/LED_circuit. There are also online calculators that will do the math for you, such as this one: http://led.linear1.org/1led.wiz.

Common SMD resistor values are listed in Table 6-2.

TABLE 6-2   Resistor Packages and Power Ratings

Maximum Power	Package Name (Imperial)
100mW	0402
	0603
125mW (1/8W)	0805
250mW (1/4W)	1206
750mW	2010
1W	2512

The package names represent the resistors' dimension in tens of mils. So the package 0402 is 40 mils by 20 mils.

For our needs, 0402 resistors would be fine, but if you've ever tried to solder a 0402 resistor by hand, you might prefer to use the much larger 0805.

Capacitors

Small-value capacitors such as the 100nF capacitor are generally available in the same sizes as the resistor sizes listed in Table 6-2; however, electrolytic capacitors use a different sizing scheme—several schemes in fact. One popular size range is the Panasonic range that is designated by a letter, e.g., Panasonic A, B, C, D, or E. When selecting the right package, you will probably need to check out component suppliers, find a capacitor of the right capacitance and voltage rating, and look at its datasheet to find its package size.

Semiconductors

Transistors and ICs also have a variety of package sizes. For example, a 555 timer is generally available as a through-hole DIL-8 (8-pin dual-in-line) package or a surface-mount SO8 (Small Outline 8-pin) package.

Transistors are also available in different SMD packages of different sizes. Generally, the bigger packages for transistors can handle larger currents. The most common SMD transistor size for small transistors is SOT-23.

Making a Surface-Mount Version

You will find an SMD version of the project in the file cycle_lamp_example_smd .fzz with the other downloads for this book. Unfortunately, the version of the 555 timer part that I selected (because it had meaningful names for its pins) only had a through-hole DIL package, whereas another of the 555 alternatives did have an SMD package. So the schematic switches to that version of the 555 chip. Ideally, you would make a copy of the well-labeled 555 part and then edit that part to add the SMD package; but we do not discuss creating and editing parts until Chapter 9.

Replace the Part Packages

Figure 6-28 shows the through-hole design, with the ground fill removed and all the traces unrouted. A quick way of removing all the traces from the design is to use the Select All Traces option on the Routing menu and then hit the DELETE key.

The next step is to click on each of the components in turn and then select an SMD package. The result is shown in Figure 6-29.

Change the C1 package to "0405 SMD." Notice that after you add it, the air wires to IC1 cross over; so rotate it through 180° to make it easier to connect.

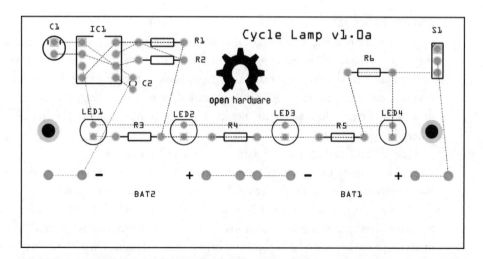

Figure 6-28   The through-hole design stripped back.

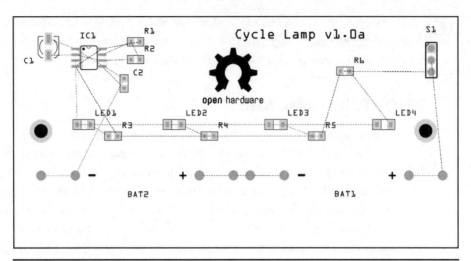

Figure 6-29   Through-hole parts replaced with SMDs.

Similarly, when IC1 is replaced with an SO8 package, it will need to be rotated 90° clockwise.

Change all the resistors and C2 to an 0805 package, rotating where necessary. Change the LEDs to the largest possible size of 1206. The PCB layout should now look something like Figure 6-30. There is now a lot of space around all the components, and we could easily reduce the size of the PCB. In this case, the lamp needs to be reasonably big so that it can be seen easily. Other board designs usually need to be as compact as possible.

Reroute the Design

When you are routing with SMDs, it is always best to start on the top layer. Since they do not have leads passing through the PCB, there is no connection to the bottom layer unless you add vias. Also if you look at the pins of IC1, you can see that our default trace size of 24 mils is actually a bit wider than the pad for the pins. So let's route the traces with a track width of 16 mils around IC1 and 24 mils where the pads are big.

The default setting for the grid is 0.05in (50 mils). SMD connections have closer spacings than their through-hole counterparts, so change the grid size to 0.01in (10 mils) by using the Set Grid Size option on the View menu.

Route all the traces around IC1 and set the trace width to 16 mils for each trace. Leave the unrouted connections that are to be connected to ground, as you are going to have ground fills top and bottom in this design. When the area around IC2 is complete, it should look like Figure 6-30.

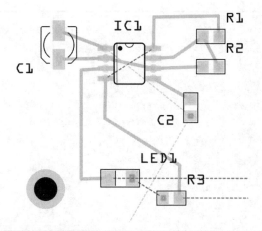

Figure 6-30 Routing around IC1.

After IC1 is routed, do as much of the routing as possible on the top layer, as that is where the SMD pads are. You can do all this routing at the default 24 mils. The connection between pins 4 and 8 of IC1 can be taken care of with a long trace from the right-hand connection of R1 round to the left connection of R6. Follow the trace through S1, R5, R4, and R3 to see how that connection works. In this circuit it will not matter that we have such a long connection. But in some designs, especially in audio amplifier or radio-frequency designs, you need to keep traces as short as possible. In that case, you would route the connection between pins 4 and 8 of IC1 directly, using vias.

The end result of this is shown in Figure 6-31.

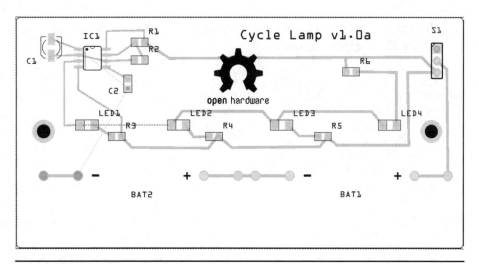

Figure 6-31 Top layer routing.

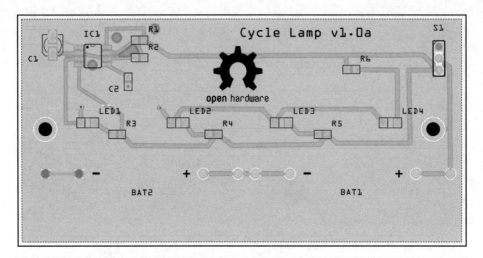

FIGURE 6-32   The completed surface-mount layout.

The only traces that remain to be routed are the ground connections that will be taken care of by the ground fills and the connection between the left-hand pad of LED1 and the left-hand pad of LED2. Route the connection between LED1 and LED2, using a pair of vias, as shown in Figure 6-32. At this point, add back the ground fills on both layers to complete the layout. To do this, you will need to first set the Layers control to allow connections on both layers.

Summary

In Chapter 7 you will learn how to take your PCB layout, turn it into a real PCB, and solder components onto it.

Fabrication

Now that we have designed a PCB, the next step is to have the PCB made for us. Although it is possible to etch your own fairly crude PCBs, the cost involved is probably higher than having the boards made by an Internet service.

Internet services are also available that take the step beyond simply making the bare PCBs and include actually soldering the components onto it. This chapter will also tell you how to put your own components onto the PCB, for both through-hole designs and surface-mount projects.

Checking the Design

As my patient woodwork teacher at school would say in response to yet another pig's ear of a dovetail joint, "*Measure twice and cut once*." Nowhere is this truer than in PCB design. If you are using a service over the Internet, then unless you are paying for a fast turnaround, you likely will have to wait a couple of weeks for your PCBs to arrive. So the last thing you want is to get the PCB back and immediately notice something wrong with it.

Before you send off the design files to have your PCB made, check the design, using the techniques outlined below, until you think that it is perfect. It probably won't be perfect, but it should work just fine and you should at least be close to making the next batch perfect.

Paper PCB

One of the quickest and easiest ways to start checking your design is simply to print a copy and then lay the components onto the paper in their correct positions. Before you print the PCB in Fritzing, select Both Layers from the Layers control and View from Above from the View control. The resulting printout will be the same size as the PCB. Figure 7-1 shows such a printout with some of the components laid on top of it.

Do the Components Fit?

The first thing to check, especially with a through-hole design, is that the parts you intend to use with the design will actually fit the pads of the PCB. There isn't going to be a problem with the common components, unless you accidentally selected the wrong package when you were laying out the PCB. But it's reassuring to check that everything will fit. You can even poke holes in the paper and push the component leads through, if you like.

There is more likely to be a problem with unusual components such as the AAA battery holder in the cycle lamp project, where the pin spacing is very specific to that particular part.

Is Everything Connected Correctly?

The second thing to check is that the design has all the necessary connections and no unnecessary connections. Since you can see the traces from both layers, a good

FIGURE 7-1   A paper prototype.

technique is to also print the schematic diagram and then, for each net on the schematic, tick off the corresponding trace or fill on the PCB.

As you gain experience and confidence in Fritzing, you will probably check less and less as you get a feel for what might be risky, which will allow you to concentrate on likely problem areas.

PDF Files

The design files that you send away to the PCB service are called *Gerber files*. These files are images in a special format that can only be viewed with a special viewer (see the next section). However, you can have Fritzing produce an equivalent set of PDF files that can be opened just as any other PDF. Checking the PDFs should be good enough, because unless there are bugs in Fritzing's Gerber file generation code, if the PDFs are correct, the Gerber files will be correct.

To generate a set of PDF files, select the menu option Export | For Production | Etchable PDF. This will open a file dialog where you will select a directory to which all the PDF files will be written. Create a new empty directory for this to make it easier to find the files.

These PDFs can also be used for do-it-yourself (DIY) PCB production using photoetching, milling machines, or toner transfer.

Figure 7-2 shows the folder containing the generated PDFs for the cycle lamp project.

There are actually 16 files, but you really don't need to check all of them. The most critical part of the design is that the copper traces are correct. So the first files to check are the top and bottom copper layers. Figure 7-3 shows the two files copper_bottom and copper_top.

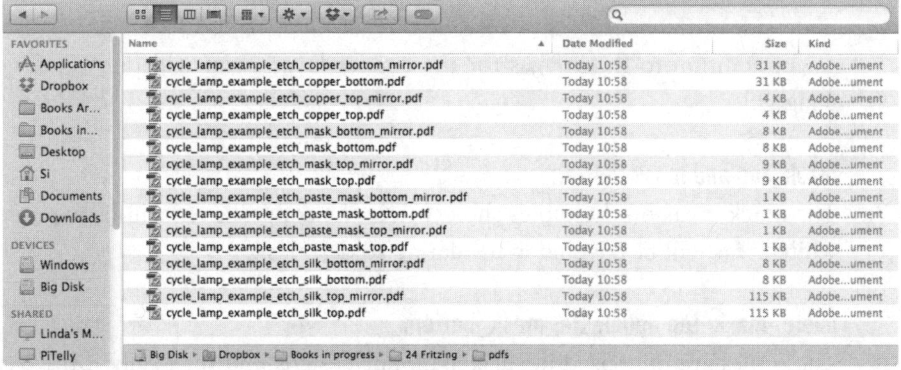

Figure 7-2 Generated PDFs for the cycle lamp project.

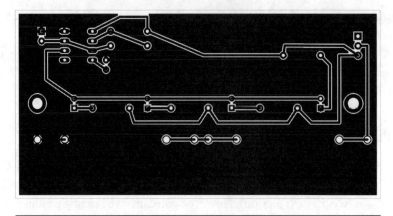

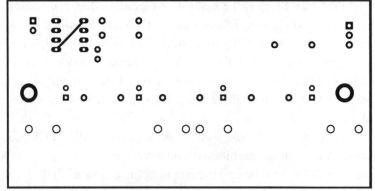

Figure 7-3   The bottom and top copper layers.

You can open these in Adobe Reader or any other PDF viewer. Printing them on thin paper will allow you to superimpose the layers by holding them up to the light. But beware; copper fills will soon eat through your printer's ink supply.

Alongside the PDF files for the view from the top of the board, there is also a parallel set of image files labeled "Mirror," which will be the view from the underside of the board.

In addition to the images for the copper layers, you will find the following image files.

mask_top and mask_bottom

The black areas on these files show where the copper layer will be exposed. The white areas will be covered in insulating solder mask.

paste_mask_top and paste_mask_bottom

These files are only used in surface-mount designs and then only if the design is going to have surface-mount components attached during manufacture, or you

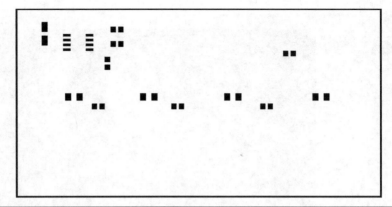

Figure 7-4   The top paste mask layer.

are buying a solder paste stencil (see the section on PCB fabrication for surface-mount technology later in this chapter). Figure 7-4 shows the top layer of this file for the surface-mount version of the cycle lamp project.

As you can see, the black areas correspond to the surface-mount pads. During the process of attaching the SMDs, a stencil will be created with the black areas cut out of a metal or mica sheet. They are then placed over the newly made PCB, and solder paste is smeared over the stencil with an applicator so that when the stencil is removed, all the pads are covered with a layer of solder paste.

silk_top and silk_bottom

The silk layers will contain the text and other markings that will appear on the PCB's solder mask layer.

Gerber Viewers

The techniques described above should give you sufficient reassurance that your design is just fine. If you need more, then you might like to view the Gerber files that will eventually be used to manufacture the PCB. To do this, you will need a special viewer such as MCN Gerber View for the Mac (www.mcn-audio.com/sharewares/) or you can use an online viewer such as www.gerber-viewer.com/.

You will need to create the files in much the same way as you exported the PDF files, only this time, select the option Extended Gerber (RS274X).

Figure 7-5 shows the MCN Gerber viewer in action. Very usefully, you can toggle the visibility of each of the files, allowing you to superimpose them on top of each other.

One layer missing from the PDF files is the drill file that shows where the holes are. This is present in the Gerber files and has the extension .TXT.

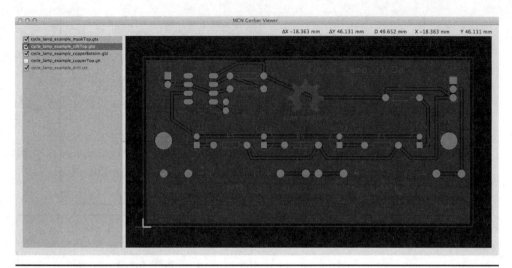

Figure 7-5   Use of a Gerber viewer.

PCB Fabrication

Now that you are happy with your design, it is time to actually make the design for real. Although you can make your own PCBs using photoetching, toner transfer, or milling machines, frankly the results will look amateurish. There is an element of craft and trial and error in all these methods, when you do them at home. So if you enjoy that kind of thing, take a look at some of the many online tutorials that cover these processes.

An interesting development is the arrival of machines such as the Voltera V-One machine (http://volterainc.com/) designed for amateur use to create PCBs on your desk as easily as 3D printing. Until such machines are cheap and readily available, you will probably be using a PCB service.

Choosing a Service

There are lots of PCB services out there, not the least being Fritzing itself. Hovering over the Fabricate button in Fritzing will even tell you how much your PCB will cost. You also have the reassurance that if your design is made using Fritzing, there should be no problem at all with design rules that you might need to check if you used another service. Using Fritzing will also help support the continuing development of this excellent free software.

The process of having your PCBs made will be much the same whether you use Fritzing or some other service. The example I use later is for the service offered by Seeed Studio, but there are new providers cropping up all the time and it can pay to shop around. You will find some providers listed in the Appendix.

Most providers will expect you to upload a zip file containing the Gerber files or sometimes to e-mail the file to them. Conveniently, most will then use this information to give you a price for the fabrication. Another advantage of using Fritzing is that you can just upload your .fzz file and they will generate the Gerber files.

Fabrication with Fritzing.org

If you plan to use the Fritzing service, you can just click on the Fabricate button at the bottom of the Fritzing window when you are on the PCB tab. Clicking on the Fabricate button will open the page http://fab.fritzing.org/fritzing-fab in your browser, where you can start the process of ordering your boards.

Click on the Submit Your Order button. This will prompt you to log in or register with Fritzing. Once you have done this and clicked on the e-mail confirmation link that will be sent, you can click on the Add Sketch button on the Web page, navigate to the .fzz file for your design, and say how many copies of the board you want.

At this point, you will be told the cost and you can go ahead and order the boards.

Fabrication with Other Services

Fritzing offers a premium service and will do a certain amount of manual checking of your design before they go ahead and make the PCBs. They are not going to check your electronic skills, but will probably spot any glaring problems with your project. There are, of course, lower-cost services that will make your boards exactly as you specify without much manual checking. Such services will not work directly from your .fzz file; instead you will need to export the Gerber design files as described earlier.

Gerber Files for Fabrication

You may have already generated your Gerber files, if you decided to view them in a Gerber viewer. If you didn't export them earlier, then export the files now. The list of files that will be generated by Fritzing is summarized in Table 7-1.

TABLE 7-1   List of Files Generated by Fritzing

File Name	File Extension	Description
contour	.gm1	This image is of the board outline. It does not include any holes on the board.
copperBottom	.gbl	The bottom layer of copper.
copperTop	.gtl	The top layer of copper.
drill	.txt	Any holes in the board.
maskBottom	.gbs	The solder mask for the bottom layer.
maskTop	.gts	The solder mask for the top layer.
pasteMaskBottom	.gbp	The paste mask for a solder paste stencil (bottom layer).
pasteMaskTop	.gtp	The paste mask for a solder paste stencil (top layer).
pnp	.txt	This file is not strictly part of the Gerber file set. It can be used by Pick 'n Place machines to automatically put SMDs onto the PCB.
silkBottom	.gbo	The text and other marks to appear on the top silkscreen layer.
silkTop	.gto	The text and other marks to appear on the bottom silkscreen layer.

Note that if your design is entirely through-hole, no pasteMask files will be generated.

Fabrication Guidelines

Each fabrication service will have its own guidelines about what you can and can't do in your PCB design. The Fritzing software uses sensible defaults for everything and produces Gerber files that are very unlikely to cause problems for any fabrication service. But, just in case, let's look at a typical set of guidelines provided by Seeed Studio for their Fusion PCB Service.

The service starts by explaining what files it will need to produce the bare PCB without any assembly. It is common for just the file extension to be used to identify the file, even though Fritzing includes a more descriptive name for each file. The descriptive name will not be a problem, but you must make sure to supply a set of files with the extensions expected. The list below is taken from Seeed Studio's PCB Order Submission Guidelines page.

```
Top Layer: pcbname.GTL
Top Solder Mask: pcbname.GTS
Top Silkscreen: pcbname.GTO
Bottom Layer: pcbname.GBL
Bottom Solder Mask: pcbname.GBS
```

```
Bottom silkscreen: pcbname.GBO
Board Outline: pcbname.GML/GKO
Drills: pcbname.TXT
```

Most of the file names coming out of Fritzing are just fine for this, and the three extension letters can be uppercase or lowercase. The only potential problem with the set of files from Fritzing is that it includes two .txt files: drill.txt and pnp.txt. The specification above specifies that the drill file should have an extension of .TXT, so delete the file pnp.txt from the files generated by Fritzing, to avoid any possible confusion. The pnp (pick and place) file is used only if the PCB is going to be populated with components.

Make sure that your folder contains a file with each of the extensions listed above and then zip the folder. This is the folder that you will upload to the service.

You should also find a long list of specifications for the PCB in the submission guidelines. This is quite long for the Seeed Studio service, so I will just pick out some of the key things to look for in Table 7-2.

TABLE 7-2   A Sample of Board Specifications

Item	Unit (mm)	Unit (mil)
Minimum trace width	0.1524	6
Minimum silkscreen text size	1	40
Drilling hole diameter (mechanical)	0.3–6.35	11.81–250.00

A minimum trace width of 6 mils is very fine; the smallest size that Fritzing provides in its drop-down list is 8 mils. The specification also specifies a minimum text height. This is unlikely to be problematic as it is almost microscopic. Note the relatively small upper size limit for drills of 6.35mm.

Fabrication with Seeed Studio

Once you have ensured that your design is not contravening any of the fabricator's rules, it's probably time to upload the Gerber files and choose any options for your PCB fabrication. Here the process is usually similar to using the Fritzing service except that you usually upload a zip archive file containing the Gerber files. Some services ask you to e-mail the files rather than upload them.

Once you have uploaded the design files as a single zip archive, the service may provide some automatic checks on the files and calculate the PCB size; or you may have to set that option manually, as is the case with Seeed Studio. Figure 7-6 shows the Seeed Studio options page, with the PCB dimension set to "5cm

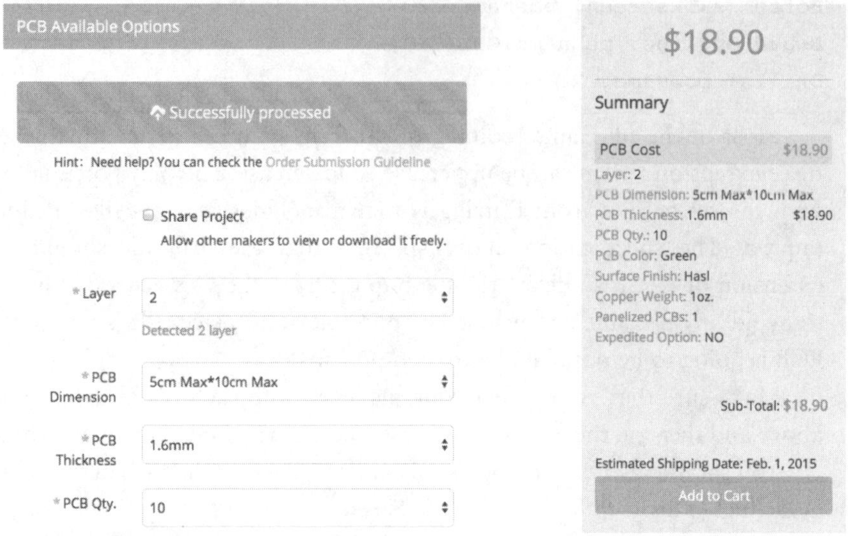

Figure 7-6   Setting options in Seeed Studio.

Max*10cm Max." The size specified here needs to be large enough to completely enclose the PCB, whatever shape it is.

Most of the options can be left at their default values.

- The default for PCB thickness is 1.6mm, and this is fine for most situations. If you have a very small or large PCB, then you may want to decrease or increase the thickness, respectively.
- The quantity defaults to 10 boards, but 5 boards is an option that saves a little money and will be more than enough for a prototype. All services including Fritzing have much lower cost per board, the more boards you order.
- Most services will offer you a choice of different colors for a little extra cost.
- The copper weight option is usually 1oz per square foot of copper, but you may get an option of 2oz for extra cost.

You may also see an option for panelizing. *Panelizing* means repeating the same PCB design on the same area of PCB material, with a groove scored almost all the way through, between each PCB, so that they can be snapped apart. The idea is that if the PCBs have a standard size of 5cm × 5cm, you could fit 10 boards 1cm × 2.5cm onto the same square, saving you a lot of money.

Some fabrication services just ban panelizing outright, and others will let you do it for an extra cost. However, Fritzing does not generate Gerber files to support panelizing, so it's a moot point.

The rest of the ordering process is just like any online retailing. You hand over your money and sometime in the future your PCBs will arrive, ready for you to assemble them, but first you are going to need some parts.

Parts and BOMs

BOM stands for Bill of Materials. This is basically a list of the parts you use in your project. Eventually, when it comes to having your design manufactured, the BOM will probably be the most significant factor in determining the production cost of your design.

As well as exporting the Gerber files and PDFs, Fritzing can export the BOM for you as an HTML page (Figure 7-7).

You will find the option to export the BOM under the File | Export menu.

The file generated will tell you all the parts you need. But if this is eventually going to become a product, you will probably want to paste its contents into a spreadsheet, so that you can add some columns indicating sources for the various components as well as their costs.

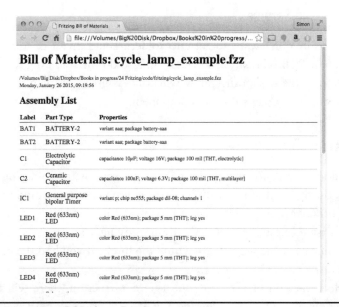

FIGURE 7-7   Exporting a BOM from Fritzing.

Finding suppliers for the parts can be very time-consuming. The big international suppliers such as Farnell, Mouser, and Digikey stock so many thousands of components that it can be hard to see the wood for the trees. However, they do allow you to get your prototyping components from one place. In fact, bulk discounts for high quantities will generally make the large suppliers quite competitive in price, especially for very low-cost components such as resistors and capacitors.

Later, when it comes to finding the cheapest source for each component, the website http://octopart.com is invaluable for finding suppliers of components. Octopart is effectively a search engine for parts suppliers. You just enter the part number that you are looking for, and the price at all the suppliers that stock it will be displayed.

Now that you have a bare PCB and a load of parts, all that remains is to solder one to the other to make your prototype.

Soldering

Hand soldering is a skill that is easy to do, but difficult to do really well unless you have a talent for it. It is, however, very satisfying to solder the parts for your project onto a fresh new PCB.

Prototyping Services

If you really don't want to get into soldering and invest in soldering equipment, then some PCB fabrication services will also do the soldering for you for a fee. In addition to the design files, they will want you to complete a BOM spreadsheet that includes sources for every component.

Tools

You do not have to spend a lot of money on tools for soldering PCBs. You can get perfectly good results with low-cost equipment. You wouldn't learn the violin on a Stradivarius, so don't get a top-end Weller or Hakko soldering station as your first soldering iron. Gradually improving your tools is one of the joys of electronic construction. Where would the fun be if you had the best of everything from day 1?

However you plan to do your own construction, there will be certain tools that you will need when soldering a PCB.

Snips and Pliers

Snips are used to cut the excess lead off components after they have been soldered. They are sharp and allow you to get close to the solder joint. They are also useful for stripping the insulation off wire. Eventually snips lose their sharpness and become blunt, especially if they are abused by being used to cut steel guitar strings. I use very cheap snips (see Figure 7-8) and then replace them as soon as they become blunt enough to be irritating to use.

Long-nosed pliers last longer and are generally a useful tool to have around. They can be used to grip insulation tightly while you strip the insulation off a wire, or for holding onto components that you are trying to desolder from a board without burning your fingers. You might find yourself desoldering a board because you may have put the wrong-value resistor in place, or you may want to swap out a component to improve the design.

Multimeter

In a perfect world, everything would work the first time that you powered it up. The reality is that life is not quite like that. A multimeter (Figure 7-9) is an essential tool that will allow you to diagnose problems with your designs.

You do not need to spend a lot of money on a multimeter. A basic entry-level multimeter costing just a few dollars will do just fine most of the time. The most important setting that you will use most of the time is DC volts on a range of 0 to 20V.

FIGURE 7-8   Snips and pliers.

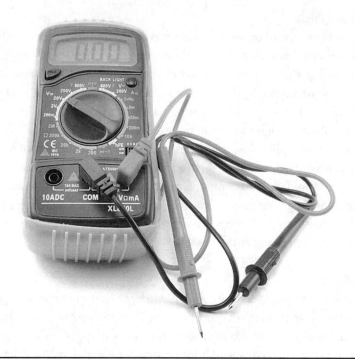

FIGURE 7-9   A multimeter.

It is also useful to have a DC current setting of up to a few hundred milliamperes and a continuity test that buzzes when the test leads are connected. Everything else is just bells and whistles that you might use once in a blue moon.

Most of the time, accuracy is pretty irrelevant too. When things go wrong, it is usually a matter of orders of magnitude. So if your multimeter indicates a current of 10mA when the current is actually 12mA, usually that is good enough. It's when the current is 100mA and you were expecting 10mA that there is a problem.

Soldering Station

While you can get by with a soldering iron that plugs directly into an AC outlet and has no way of adjusting the temperature, it is worth spending a few extra dollars on something that is thermostatically controlled and can accept fine-point tips (Figure 7-10). Make sure that you avoid anything advertised as being suitable for plumbing use.

When you are buying your soldering station, bear in mind that eventually the tips (also called *bits*) will need to be replaced. So make sure that replacements will remain available, or buy them when you buy the iron. With the trend for components to get smaller and smaller, you will probably want a tip of perhaps

Figure 7-10 A low-cost soldering station.

2mm. There are many different shapes of tip, and it is a matter of personal preference. Many people prefer a chisel-shaped tip. A simple conical tip is another popular choice.

If you plan to use lead-free solder, then temperature control is a must. You can get away with a simple low-cost soldering iron if you are using solder with lead in it, as this is just much easier to work with (see the next section).

WARNING *It should go without saying that soldering irons get hot enough to burn your skin. So be very careful and always put the soldering iron back into its holder, as soon as you have finished with it. Do not leave it on the desk to roll off, triggering the automatic instinct to try and catch it, when inevitably its lead gets snagged and it falls off the desk.*

Soldering also produces fumes from the rosin flux. It is a good idea to solder next to an open window, or use a fume extractor.

Solder

Traditionally, solder (Figure 7-11) has been made from tin and lead. Usually, this is in the proportion of 60 percent tin and 40 percent lead. The solder looks like a solid metal wire, but actually will normally have a core of flux rosin that helps the lead to flow when it melts. Legislation on the use of toxic chemicals has caused a reduction in the use of lead-based solder in favor of lead-free solder.

Figure 7-11   A reel of 0.7mm tin, lead solder.

This type of solder is an alloy of tin, silver, copper, and small amounts of other metals. It looks like lead solder and still has a rosin core, but is more brittle than lead solder and has a melting temperature of about 200°C (392°F) versus about 190°C (374°F) for leaded solder.

The differences do not end there. Many people find lead-free solder much harder to work with. It does not flow as easily as lead solder.

Lead solder is still widely available, and unless you are producing a product that you are going to sell, then it is really a matter of personal preference which type of solder you use. I know electronics enthusiasts who have a roll of lead-free solder that they use most of the time and then a roll of "the good stuff" (lead solder) when they have something tricky to solder.

Whatever type of solder you use, you will have another choice to make of what gauge of solder to buy. Two popular sizes are 0.7mm and 1.2mm diameter. Use 0.7mm or similar, as it is much easier to use when there are IC leads close together. If you need to solder some large terminal, you will find yourself feeding in quite a length of the narrow solder to deliver the required amount, but this is not really a problem.

Desoldering Braid

Desoldering braid (Figure 7-12) is not an essential tool for soldering; however, it can come in very handy from time to time. As well as its primary use for "unsoldering" components, it is also great for mopping up excess solder, especially when hand-soldering surface-mount devices.

FIGURE 7-12   Desoldering braid.

The braid is made of copper impregnated with flux that encourages the solder to flow. So when you place it between the pad that you want to remove the solder from and the soldering iron tip, it soaks up the solder as a sponge would. Once this is done, that section of braid cannot be reused; you snip it off and throw it away.

Tip Cleaner

When you solder, it is very important that the tip of the iron be clean, or you will end up with blobs of solder that do not make a good joint. There are two methods of cleaning the tip, both used with the soldering iron hot. One is to use a damp sponge, and many soldering stations include a sponge holder. The other is to use a container of brass wool, rather like a scouring pad (Figure 7-13).

FIGURE 7-13   Brass soldering iron tip cleaner.

The only real advantage of using a damp sponge is that it makes a great hissing noise as the hot tip of the iron is rubbed across it. It does, however, suffer from a number of disadvantages:

- The thermal shock of cooling the tip quickly as it comes into contact with the wet sponge will shorten the life of the tip.
- You have to keep wetting the sponge and need a supply of water.

Tools for Surface-Mount Devices

When attempting surface-mount soldering by hand, you will probably need all the equipment that we have just described for through-hole soldering. In fact, you can get away with just using regular soldering equipment if you select the larger SMDs. However, there are a number of special items that make surface-mount soldering simpler.

Hot Air Gun

A hot air gun (Figure 7-14) has interchangeable nozzles of different sizes and allows you to deliver a stream of hot air to an area of a circuit board.

You can normally set both the temperature of this air and the flow rate. You will need to control the flow rate, because many SMT components are so small that they can easily be blown away by the pressure of air from a hot air gun.

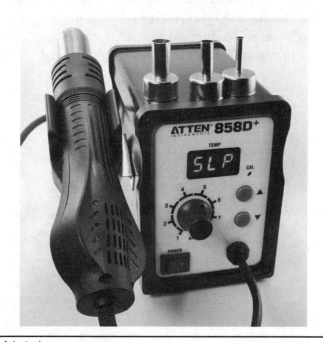

FIGURE 7-14   A hot air gun.

Solder Paste

When soldering surface-mount devices, with care you can use regular solder. However, if you are using a hot air gun or a reflow oven, then you will need to use solder paste (Figure 7-15). Solder paste is available in both lead-based and lead-free varieties and suffers the same pros and cons as those variants of regular solder.

Solder paste is made from microscopic spheres of solder in a suspension of flux. For industrial use it is supplied in tubs; for small-scale use, you can buy it in syringes ready for hand use. Solder paste should be kept in your fridge but warmed up to room temperature when you are ready to use it.

WARNING *Solder paste is a liquid. If you are using lead-based solder paste, it will easily find its way into the pores of your skin if you get it on your fingers. So, always wear latex gloves, if, like me, you are a bit messy and are likely to get it on your fingers.*

Tweezers

To be able to pick up and place SMDs onto a board, you will need tweezers (Figure 7-16).

The tweezers should have a fine point and most importantly be nonmagnetic. If they are even slightly magnetic, then SMDs will probably stick to them, as many contain ferrous metals.

FIGURE 7-15   A syringe of solder paste.

FIGURE 7-16   Nonmagnetic tweezers.

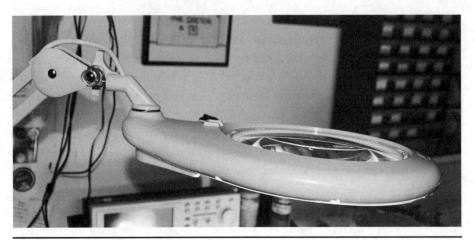

FIGURE 7-17   A magnifying work lamp.

Magnifier

It can be really hard to see what you are doing when you are working with SMD. A large magnifying work lamp, such as that shown in Figure 7-17, can be a great help.

These devices have a lighting ring around the lens that evenly illuminates the board you are working on. Since you are looking through the lens with both eyes, all-important depth perception is preserved.

Some people take this a stage further and use a binocular microscope. These are available specifically for working on circuit boards, and a zoom version will allow you both to work on boards and inspect them very closely for any problems. You should look for something that magnifies between 5 and 20 times.

Reflow Oven

When you are developing single boards, hand soldering works okay. It is a little tedious and time-consuming, but can be done. The professional way to attach SMD components to a board is to use a reflow oven.

The basic idea is that you put solder paste on the pads of the board, place the components onto the pads, and then bake the entire board in an oven to melt the solder paste and attach the components. We will see how to do this in a later section.

Commercial reflow ovens are quite expensive, but many people make their own by using low-cost toaster ovens, such as the modified device of the author's shown in Figure 7-18.

WARNING *These types of toaster ovens are often called "fire starters" and for good reason. They are very simple designs, with little in the way of thermal*

Figure 7-18   A modified toaster oven.

insulation. That means that they get very hot, and if you modify them, they can become even more dangerous.

If you decide to make one of these, never leave it unattended or anywhere near anything that could burn.

The model shown in Figure 7-18 has been modified to replace the thermostat with a proportional power control module, and a digital thermometer has been added to allow the necessary accurate monitoring and control of temperature.

Through-Hole Soldering

Having explored the various tools we will need, we start by learning how to solder through-hole PCBs. It is a good idea to try to follow these instructions on a PCB. You may wish to order one of the PCB designs from earlier in the book, such as the cycle lamp example.

Through-Hole Soldering Step by Step

First turn on your soldering iron and set the temperature. You will find conflicting advice for temperatures to use, but I set my soldering iron to 280°C (536°F) for lead-based solder and 310°C (590°F) for lead-free solder. Once you get comfortable with soldering, you will probably want to work at a higher temperature where the solder melts a bit more quickly. The higher temperature will not damage the components as long as you work quickly.

When the soldering iron is up to temperature, clean it on the damp sponge or brass tip cleaner. Once cleaned, it should look bright and silvery.

The key to soldering is not to heat the solder, but rather to use the soldering iron to heat the place where you want to solder and then feed solder onto that junction so that it melts and flows over the pad and the component lead.

First push the component leads through their holes and turn the PCB on its back (Figure 7-19a); then hold the tip of the soldering iron to the junction of the pad and the lead. Next, feed solder into the joint so that it flows all around the lead and covers the pad (Figure 7-19b). Once it has flowed all around, you can stop adding solder and move the tip away. You do not want the pad to be heaped high with solder. The solder should ideally form a nice mountain shape around the lead (Figure 7-19c). Finally, you can snip off the excess lead (Figures 7-19d and e).

a

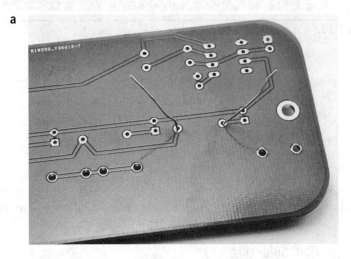

b

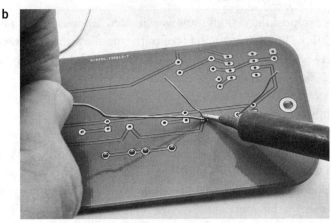

Figure 7-19   Soldering a resistor.

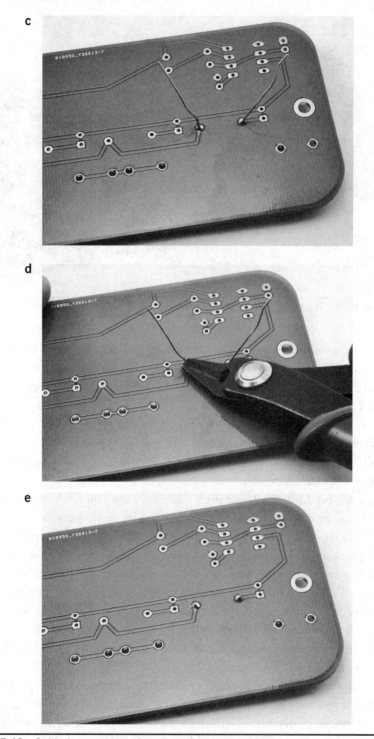

Figure 7-19 Soldering a resistor (*continued*).

Figure 7-20   The cycle lamp example soldered.

When soldering components onto a PCB, you can make life much easier for yourself if you start with the components that lie closest to the surface of the board. That way, when you turn the board on its back, the weight of the board will keep the components pressed against the board.

Figure 7-20 shows the PCB for the cycle lamp project with all the components except the battery holders attached.

If you make a mistake and find yourself needing to desolder a joint, then Figure 7-21 shows the steps you should take.

If the board has not just been soldered, then you might find that it is quite hard. So heating it and adding a bit of solder could actually make it easier to desolder the component. After optionally resoldering the joint as described above, you place an unused end of the soldering braid against the joint (Figure 7-21a) and then press it down onto the solder pad, using the tip of the soldering iron (Figure 7-21b). As the solder melts, it should be drawn into the solder braid. You will be lucky to draw off all the solder in one go, so most likely you will need to snip off the now solder-covered end of the braid and repeat the process of pressing it against the joint. Eventually, you should have most of the solder removed, and the joint will look like Figure 7-21c.

Repeat this for the other lead or leads of the component. If you are very lucky, you will just be able to wait until the component has cooled and then gently pull the component back through the hole from the top. However, it is more likely that there will still be a little solder holding the component in place. If this is the case,

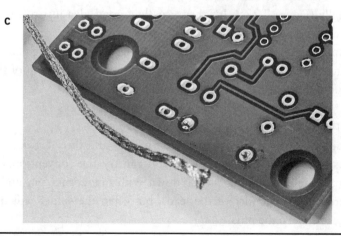

FIGURE 7-21   Desoldering.

then hold one component lead from the top side with long-nosed pliers and heat the pad from underneath while exerting a gentle pull on the lead to pull it back through. If you cannot get to the lead and you don't mind breaking the component, then snip it in half, or snip the lead off on the top to make it easier to pull through.

All this means that you are likely to be heating the board for a while, which can make it look scruffy or even damage the board. Pulling the lead through with force is also likely to damage the PCB, and ultimately too much heat will eventually cause the pad to separate from the board.

SMD Hand Soldering

As long as the SMDs you use are at the larger end of the size scale and have pins on the edges of the devices rather than on the underside, they can be soldered relatively easily with a regular soldering iron. The main problem is that they are so light and since they do not have leads projecting through holes, there is nothing to hold them in place when you try to solder them.

Soldering Two- and Three-Leg Components

The sequence of steps to solder a 1206 resistor into place is shown in Figure 7-22.

This is one of those situations that would really benefit from having three hands—one to hold the soldering iron, one to hold tweezers to keep the SMD in place, and one to apply the solder. If you do not have three hands, then a good trick is to place a small mound of solder on one of the pads (Figure 7-22a) and then, while holding the SMD in place with the tip of your tweezers, press its lead into the little solder mound (Figure 7-22b) with the tip of your iron. The SMD will now stay in place without the need of tweezers as you solder the other end normally (Figure 7-22c). It's then usually a good idea just to touch the first end with the iron and a little solder, just to freshen it up.

An alternative technique is to place solder paste onto the pads, hold the SMD in place with the tip of your tweezers, and then touch the tip of the iron to each lead until the solder paste beneath melts.

Soldering IC Packages

The approach above works just fine for two- and three-leg devices, but when it comes to ICs, this can be trickier. You can try the conventional soldering iron approach above, pinning the IC down with one corner pin and then carefully soldering the remainder of the leads, but often the solder will make unwanted bridges between the pins.

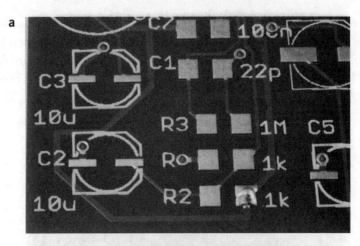

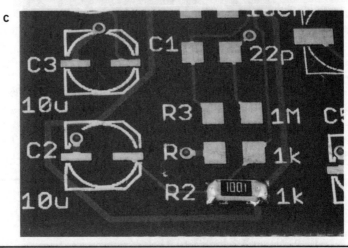

Figure 7-22   Using a soldering iron on an SMD resistor.

A good trick is not to worry too much about these bridges, but when you have finished soldering, lay desoldering braid along the row of pins and heat along the whole length to remove the excess solder. This is shown in Figure 7-23. This will still leave enough solder under the IC pins to make good connections.

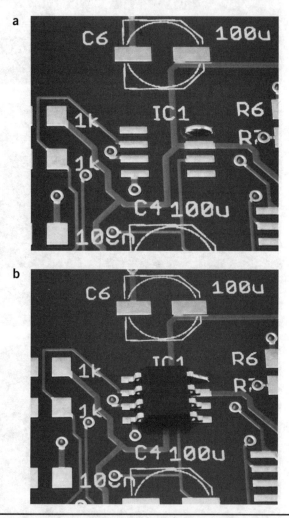

FIGURE 7-23 Hand soldering an SMD IC.

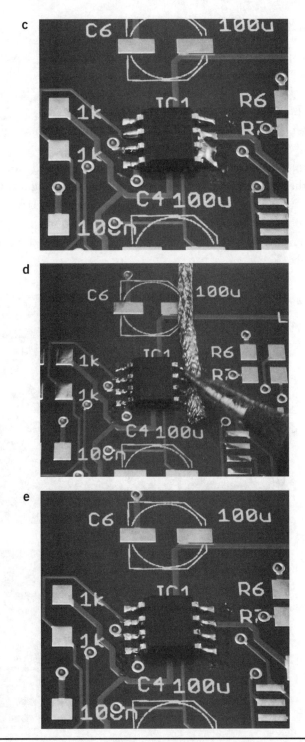

FIGURE 7-23   Hand soldering an SMD IC (*continued*).

SMD with Hot Air Gun

Generally speaking, it is much easier to solder SMD devices with solder paste and a hot air gun than with a soldering iron.

Soldering Two- and Three-Legged Components

The steps for soldering with solder paste and a hot air gun are illustrated in Figure 7-24.

First, place a small amount of solder paste on each pad. You can squeeze it out through the syringe needle. Use clean gooey paste, wiping away any crusty paste from the end of the syringe before you start (Figure 7-24a). Occasionally, I find a wooden toothpick useful for spreading the paste around a bit. Do not worry if it is a little untidy; when it melts, surface tension will cause it to pull back onto the solder pad.

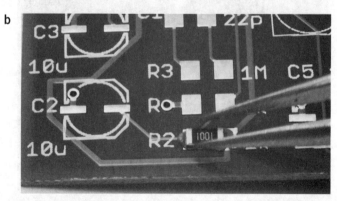

Figure 7-24   Soldering with hot air gun.

Figure 7-24   Soldering with hot air gun (*continued*).

Next, using your tweezers, place the component onto the pads (Figure 7-24b). Now fit a small nozzle onto your hot air gun, set the temperature to 280°C (536°F) for lead-based solder and 310°C (590°F) for lead-free solder, and set the flow rate to perhaps one-quarter of full power. When the air gun is up to temperature, pin the component down with the tip of your tweezers and then play the air gun over the component and its leads until the solder paste melts (Figure 7-24c). When the solder melts, you will see it change from dull gray to shiny silver and see it spread across the pad. While still pinning down the component, put the hot air gun safely back on its stand; after a second or two, when you are sure the solder has set, let go with the tweezers.

If you do not hold the component in place with tweezers, then even at very low airflows the hot air gun will probably blow the component out of position.

Soldering IC Packages
Soldering ICs is very similar to the process just described. The only real difference is that you may well want to clean up the connections using soldering braid, as shown in Figure 7-23.

Packages with Hidden Connections

Some IC packages have inaccessible components on their underside. To solder these, put paste on the pads as usual, and then while you hold the IC in place with the tip of your tweezers, place the hot air gun over the whole IC, until you feel that the solder has melted. If you do this for too long, you may damage the chip.

Using a Reflow Oven

By far the quickest way to solder an SMD board of any complexity is to use a reflow oven. It has the advantage that you only have to place the components on top of the solder-pasted pads. Once that is done, the whole board is cooked in the reflow oven, soldering all the components in one go. What is more, once you gain confidence, you can cook a whole batch of PCBs in one go. In any case, the "cooking" process only takes a couple of minutes.

If you have a board that contains both surface-mount and through-hole components, then solder the surface-mount components first.

Get Everything Together

Solder paste will dry out after maybe half an hour, so before you do anything, make sure that you have all the components that you need and that you know exactly how they will fit onto the PCB (which way around LEDs, etc., are to be placed). I find it useful to actually put the components onto the PCB without any paste just to make sure that I have everything I need. I have the board on a paper printout of the PCB, so that when I am sure I have everything in place, I can move everything off to one side, keeping the same relative positions of the components.

Another approach is to use labeled bottlecaps or even just circles on a sheet of paper labeled with the component's part and/or value. Just don't sneeze on your SMDs or they will be scattered far and wide.

Applying Solder Paste

The low-tech way of applying solder paste is the same as we described earlier when we looked at using a hot air gun. Simply go around the board, adding a little blob of solder paste onto every pad on the PCB by using the syringe dispenser. This is the time-consuming part.

The alternative to using a syringe is to use a stencil. Many PCB manufacturing services (for a small extra fee) will also supply you with a stencil. This can be made of thin steel, Mylar, or other materials, and it is placed over the PCB. It masks out most of the PCB surface except for the areas where solder paste needs to be deposited. You then place some solder paste on the stencil and "squeegee" the solder paste into all the "holes" in the mask. The excess solder paste is then scraped up and the template removed, leaving solder paste on all the pads.

You can also make your own stencils. If you search the Internet, you will find various DIY techniques for doing this by using laser cutters or vinyl cutters or even transferring toner onto a cut-up aluminum soft drink can that is then put in acid to dissolve away the holes and make the template.

In this example, we are applying the solder paste by hand to the fairly densely packed PCB that has a wide variety of different component types. When the solder paste has been applied, the board will look something like Figure 7-25.

Note what a poor job the author has done in supplying the solder paste evenly. You should aim to be neater than this, but even with this level of messiness, it is still likely to work.

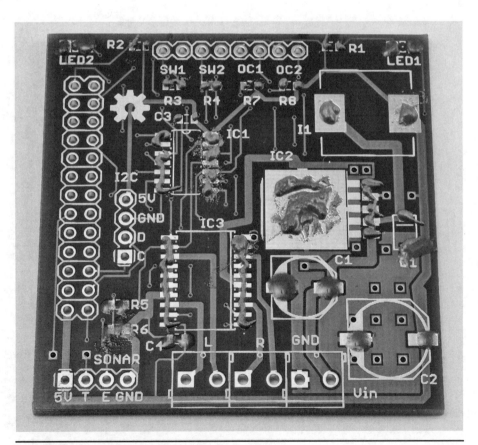

FIGURE 7-25 The board prepared with solder paste.

Populating the Board

Starting with the smallest, lowest-lying components at the far end of the board, place them on the pads, using tweezers, until all the components are placed (Figure 7-26).

The board is now ready for cooking.

Baking the Boards

If you have a proper reflow oven, then baking the boards is pretty much as simple as putting them in and pressing a button. Behind the scenes, there is some fairly careful temperature control going on.

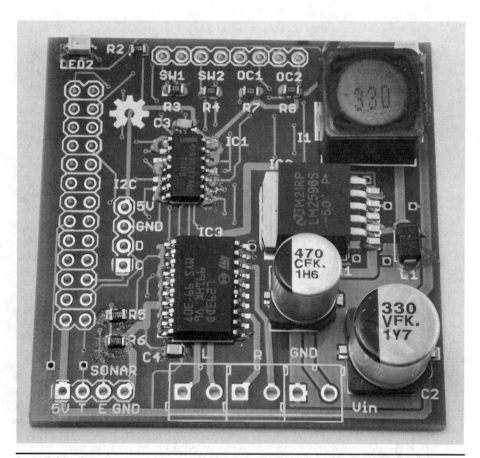

FIGURE 7-26 The populated board.

Solder paste requires a certain profile of changing temperature over time for it to do a good job of soldering components in a reflow oven. It has to go through four distinct stages:

- **Preheat:** Activate the flux.
- **Soak:** Warm the whole board to just below the solder melting point.
- **Spike:** Do this as fast as you can above the melting point to reflow the board.
- **Cool:** Cool everything before the components and board are damaged.

Each of these stages has precise temperatures and timings associated with it. A commercial reflow oven will allow you to select from preset profiles to match the paste you are using and control all aspects of the heating. Figure 7-27 shows the temperature profile of some leaded solder paste, and Figure 7-28 shows the board in the author's homemade oven.

Using a homemade device like this will never be as reliable as using a professional oven, but it is fine for prototyping. Making one of these is dangerous and should be undertaken only if you really know what you are doing. You can find instructions for doing this at the following Web pages:

- www.sparkfun.com/tutorials/60
- www.freetronics.com/pages/surface-mount-soldering-with-a-toaster-oven #.Us_cUGRdVyF

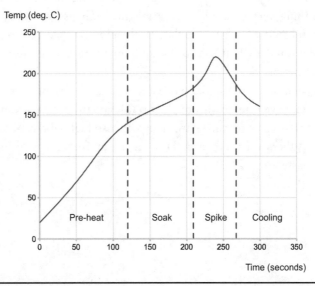

FIGURE 7-27 Leaded solder paste reflow profile.

Figure 7-28 Cooking the board.

Both of these tutorials describe how to manage the temperature by hand, without the need for a complex controller. The final board is shown in Figure 7-29. You can see that the LED has moved a little. This is an effect of my cavalier application of solder paste, but can easily be corrected with a little hand soldering.

Before you power up any board that you have made, you need to go over the whole thing very carefully with a magnifying glass, to check that there are no accidental solder bridges and that all the pins are soldered to pads. Be especially careful around SMD ICs. You can mop up excess solder causing bridges by using desoldering braid.

Depending on the solder paste that you used, you may also find little patches of flux and even tiny balls of solder on the board. I have a soft toothbrush that I use just to brush over the board. This will also highlight any loose components by brushing them off the board. This can be improved if you look for "no-clean" solder paste.

FIGURE 7-29   The final board.

Summary

In this chapter, we have learned how to get a PCB made for us and then to solder the parts onto it. So now you should be familiar with the process of creating a fully populated PCB. In Chapter 8, we will look at ways of using Fritzing with the Arduino and other popular microcontroller platforms.

Fritzing Arduino (and Other Boards)

Fritzing is a very accessible tool. It was designed from the outset as a tool for hobbyists rather than professional electrical engineers. As such, it is often used as a means of documenting and designing projects that use an Arduino, Raspberry Pi, BeagleBone Black, or other low-cost controller boards that are popular with amateurs and inventors.

What's more, it is also easy to use Fritzing to design Arduino shields. They are plug-in boards that piggyback the popular Arduino microcontroller development boards.

Breadboarding with Microcontroller Boards

Fritzing includes parts in the parts bin for a wide range of Arduino boards as well as for the Photon and the Raspberry Pi and BeagleBone single-board computers. If you are planning a project based on one of these boards, then drawing out a neat breadboard diagram is a great way of documenting the design.

Arduino

In a later section you will see how easy it is to use Fritzing to create an Arduino Shield, an add-on board that fits over the top of Arduino into its header sockets. But in this section, I concentrate on creating breadboard designs with various types of Arduino.

It used to be that if you wanted to use a microcontroller in a project, you would need to master some tricky programming and probably a Windows-only development environment to program the microcontroller. What's more, you

165

would have to spend time with datasheets understanding just how the device you wanted to use worked, setting fuses and using specialized programming hardware. This made it barely worthwhile for hobbyist users who just wanted something quick and easy to control some electronics.

The Arduino has revolutionized the use of microcontrollers. This small board (in its most popular form, the Arduino Uno) is little more than an ATmega328 microcontroller chip and some extra supporting components, including a power supply, USB interface, and crystal. It has an easy-to-use C library and development platform that will run on Windows, Mac, and Linux. You can then program the board through its USB interface. The USB interface can also be used for communication between the Arduino and a regular computer.

If you want to learn more about the Arduino and how to program it, you might like to read the book *Programming Arduino: Getting Started with Sketches*, also by Simon Monk.

Arduino Uno R3

The Arduino Uno is the most common of the Arduino boards; of these, the latest incarnation is the R3 (revision 3). Fritzing includes an Arduino R3 board. You will probably want to use a breadboard with the Arduino, so when both are added, you can start joining things with virtual jumper wires.

Figure 8-1 shows the Arduino design that you first saw as an example in Chapter 3.

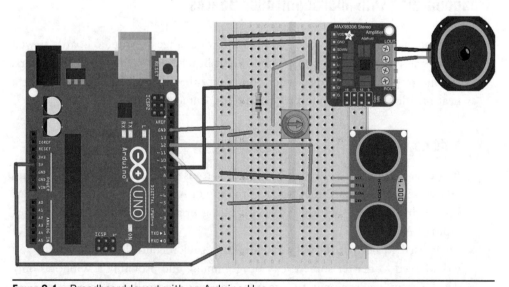

FIGURE 8-1 Breadboard layout with an Arduino Uno.

The Arduino has a useful regulated power supply that can provide both 5V up to 500mA and 3.3V up to 40mA to components on the breadboard. It is therefore quite common to connect power lines from Arduino GND and 5V to the supply rails on the breadboard.

It is a good idea to color-code the wires and then use the same colored wires when you actually connect things between the breadboard and Arduino.

Having the breadboard and Arduino side by side can cause wires to become detached when you move things about. You can buy plastic holders that have one section for the Arduino and a second section for the breadboard. This keeps the whole project together.

Arduino Pro-Mini and Nano

Several of the Arduino models are actually designed to be breadboard-ready and will fit onto solderless breadboard. The Arduino Nano is essentially an Arduino Leonardo (the same pin layout as the Arduino Uno, but a different microcontroller) that has been shrunk to fit onto breadboard.

The Arduino Pro-Mini is an altogether simpler device without a USB interface. Instead, a separate USB adapter is used when programming the Pro-Mini. This avoids the waste of having a USB interface built into the board that is required only during development.

Figure 8-2 shows an Arduino Nano and Pro-Mini with a USB programming adapter attached; Figure 8-3 shows an Arduino Nano on breadboard connected to an LED.

FIGURE 8-2 An Arduino Nano and Pro-Mini with USB adapter attached.

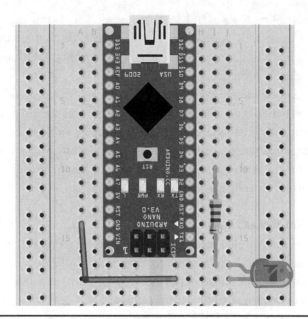

Figure 8-3   An Arduino Nano on breadboard with an LED.

Raspberry Pi and BeagleBone Black

Fritzing also has parts for the Raspberry Pi and BeagleBone Black, which you can use in just the same way as an Arduino Uno. At the time of writing, the new Raspberry Pi models B+ and A+ are not yet available as parts, but the older A and B models are available.

Protoshield Design

Once you have prototyped an Arduino-based project on breadboard, it can be useful to build something a little more solid before you take the step of designing PCBs. Arduino Protoshields are a good way of soldering a design without needing a custom PCB.

An Arduino protoshield is a shield that plugs on top of an Arduino and contains a large prototyping area. You can push through-hole components through the holes on the top of the board and then solder them together on the underside. Figure 8-4 shows an example of a protoshield project mounted on an Arduino Uno.

Fritzing includes a Protoshield part. From the schematic view, the protoshield looks just like an Arduino (Figure 8-5a), and on the Breadboard view, you can treat the board the same as breadboard (Figure 8-5b).

Figure 8-4 An Arduino protoshield project.

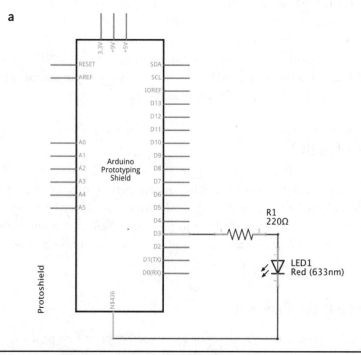

Figure 8-5 Using an Arduino Protoshield in Fritzing.

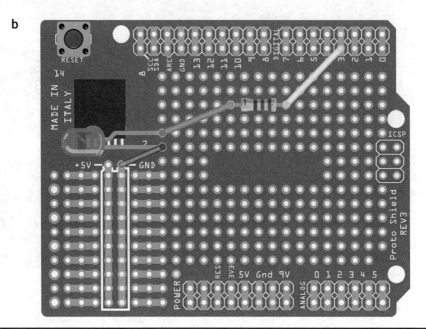

FIGURE 8-5   Using an Arduino Protoshield in Fritzing (*continued*).

On the Breadboard view, each pad is not normally connected to any other pad and just offers a secure way of anchoring the part. You will notice that some of the pads are connected; for example, the two columns labeled +5V and GND on the left of the board are connected to the Arduino 5V and GND connections using PCB tracks on the protoshield. This serves a similar purpose to the columns of holes down the sides of solderless breadboard.

Off-Board Arduino

Although the Arduino is a fantastic tool for putting together a prototype, it does not usually make sense for an Arduino board (even a Nano or Mini) to be included in a final product. When it comes to the final product, you will want to take the parts from the Arduino that you need for your invention and add in the extra parts that were connected to the Arduino.

Removing the ATmega IC

A good approach to a project such as this might be to first prototype your system on breadboard with an Arduino Uno. Then once it is all working and you have the

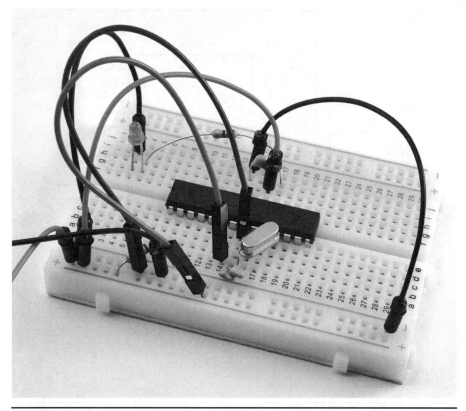

Figure 8-6   Basic off-board Arduino.

software working, you can carefully remove the microcontroller IC from its socket on the Arduino Uno and place the programmed chip onto the breadboard, as shown in Figure 8-6.

When you remove the ATmega chip, carefully ease the chip up out of the IC socket with a screwdriver, lifting each end in turn until it comes out of the socket easily. Be careful, as it is quite easy to badly bend the pins that may then break off.

Note that there are actually two versions of the Arduino Uno, the Uno and the Uno SMD. The Uno SMD costs less, but you cannot remove the microcontroller, so it is not suitable for taking your designs of the Arduino onto breadboard.

If you needed to reprogram the microcontroller, you would need to take it off the breadboard and pop it back into the Arduino Uno for reprogramming. But make sure that you put it back the right way around. Pin 1 of the chip (indicated by a dot next to the pin) goes to the end of the Uno farthest from the USB socket.

Figure 8-7 shows the schematic diagram, breadboard layout, and PCB layout for the off-board Arduino. You can find the Fritzing file for this with the downloads

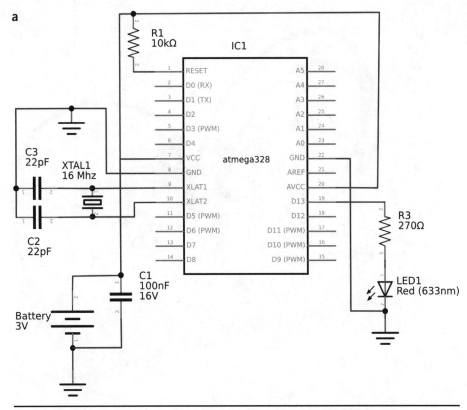

Figure 8-7   Off-board Arduino schematic, breadboard layout, and PCB layout.

for the book in the file ch_08_Arduino_baseline.fzz. All the downloads for the book are available from the author's website at www.simonmonk.org.

There is no voltage regulator, but the ATmega microcontroller IC will operate from 3 to 5.5V, so you can power a circuit such as this with two or three AA batteries. The other components included are the quartz crystal, capacitors, a couple of resistors, and an LED.

Arduino Bootloader

The ATmega microcontrollers used in Arduino boards are preprogrammed with something called a *bootloader*. The bootloader is what allows Arduinos to be programmed over USB or using a USB to serial adapter.

If you buy an ATmega328 IC from a manufacturer, it is unlikely to come preprogrammed with the bootloader, and special programming hardware is needed to install the bootloader. However, if you search the Internet, you will find a number of sources for ATmega328s that have the bootloader installed for a little

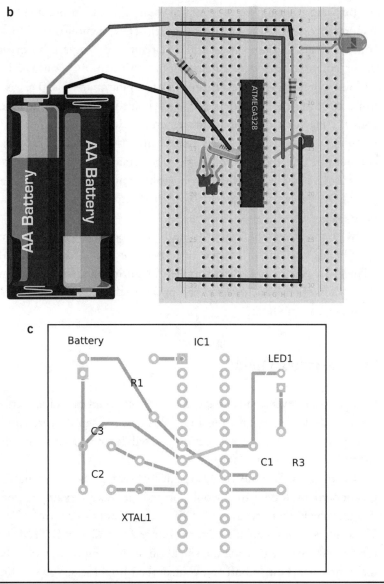

Figure 8-7 Off-board Arduino schematic, breadboard layout, and PCB layout (*continued*).

extra cost. In fact, Shrimping.it, described next, can supply the ATmega328 with bootloader.

Shrimping.it

Taking the microcontroller off an Arduino reveals just how little of the Arduino you really need to make a microcontroller-based project.

Building a Shrimp on breadboard is an approach promoted by @ShrimpingIt (described online at www.shrimping.it). They demonstrate how easy it is to construct an Arduino-compatible circuit on solderless breadboard. The documentation provided guides people through the construction of various Arduino-compatible projects using prototyping materials. This is cheaper than using an official Arduino board, and it also offers the educational value of constructing the whole circuit from scratch. It is an open procurement project, meaning build designs are freely shared and instructions are provided to buy bulk components directly from wholesalers. For smaller volumes or for reduced hassle, low-cost kits can be purchased from @ShrimpingIt directly. This is a great way of getting everything you need for your first experiments with off-board Arduino prototyping.

With the Shrimping.it approach, the microcontroller is programmed through a USB adapter similar to the one used for the Arduino Pro-Mini; and the ATmega328 microcontrollers supplied by Shrimping.it are preprogrammed with the Arduino bootloader, so you can program them as if they were Arduino Unos from the Arduino IDE.

Making an Arduino Shield

Sometimes an invention is designed for other inventors and makers. For example, you may decide to design an Arduino shield to be manufactured and sold to other makers and inventors. In this section you will learn how to create an LED lighting controller as an Arduino shield.

Creating an Arduino shield design in Fritzing leads very naturally from the breadboard prototype that would precede it. As an example, you could easily design a shield that uses three MOSFET transistors to control 12V LED lighting. These are not your regular LEDs, but rather 2 to 5W, 12V DC LED lamps intended for domestic lighting. Using the shield on an Arduino, you would be able to turn these three lights on and off and control their brightness. You will find an Arduino test sketch for this with the downloads for the book at www.simonmonk.org along with the Fritzing file for this project, which is called ch_08_lighting_shield.fzz.

Breadboard Layout

In this project, begin with the breadboard layout. It makes sense to start with just one of the three lighting channels and then duplicate the other two when the first channel is working. Figure 8-8 shows the breadboard layout for just one channel.

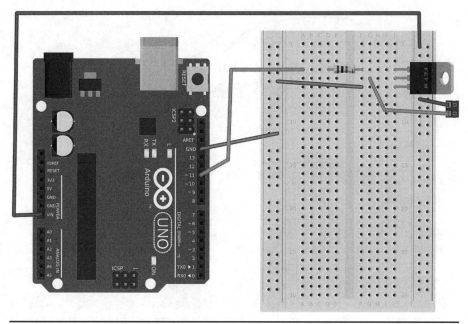

FIGURE 8-8   Breadboard design for one channel.

The Arduino will be supplied with 12V DC through its DC jack. The 12V will then be available at the V_{in} pin, which can supply the power to the LED lamps, each of which will be switched by a MOSFET. The design is actually very similar to the motor control design in Chapter 4.

The components are all from the CORE parts bin, except the screw terminal just below the MOSFET. That terminal can be found by searching for Screw Terminal and selecting the option with 0.2in pin spacing. Eventually, the lamp will be attached to the screw terminal on the PCB, but during testing on breadboard, a combination of male-to-male jumper wires and alligator leads will work just fine (see Figure 8-9).

Add the Other Channels

Build the first channel on breadboard and test it out, before you add the other two channels to the design by selecting the MOSFET, the screw terminal, the resistor, and the leads to the resistor and screw terminal and using the Duplicate option on the right-click menu.

Reposition the groups of components, add in the wires to Arduino pins D10 and D9, and you should end up with a breadboard layout that looks like Figure 8-10.

Figure 8-9 Prototyping the lighting controller on breadboard.

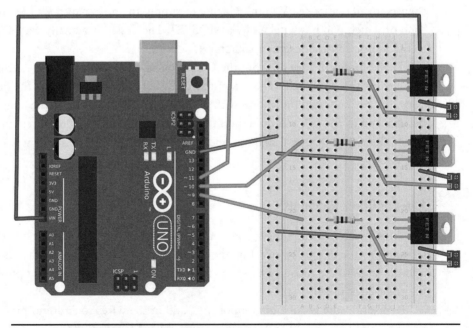

Figure 8-10 Final breadboard layout for the shield.

Test all three channels (just in case), and then you are ready to start designing the PCB.

Layout of the PCB

When you first switch over to the PCB tab, you will see something like Figure 8-11. Note that I have changed the background to white.

The components are all over the place, and although there is the outline of an Arduino board, this represents the Arduino itself, not the PCB for the shield. The PCB for the shield is currently rectangular, so select the PCB and then, using the Inspector, change the board shape to "Arduino Shield."

Now line up the Arduino itself and the shield, so that the pins are in the right place. You will also need to separate out the components that you duplicated earlier that will be stacked on top of one another.

Finally, drag all the parts onto the PCB, and do the usual rotating and label tidying, to produce component positions as shown in Figure 8-12. Start by positioning the screw terminals, as these need to be easily accessible from the side of the board.

When positioning the through-hole parts on the shield, you need to think about the underside of the shield and what's on Arduino below. The screw terminals, in particular, have fairly long through-hole leads that could touch something metallic on the Arduino below. Making a paper prototype for a project like this is always worthwhile.

The nets on this design fall into one of two categories. There are power nets, which will be carrying significant but not huge currents (6W at 12V is 0.5A), and then there are control nets that turn the transistors on and off, through which very small currents will flow. The current will be at most a few tens of milliamperes. That is, after all, the whole point of using a MOSFET as a switch.

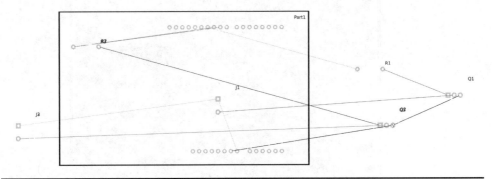

Figure 8-11   The initial PCB view for the Arduino shield project.

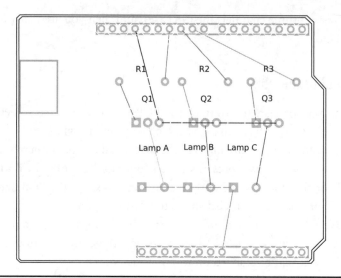

FIGURE 8-12   Parts in position and ready for routing.

It makes sense to route the control signals in standard 24 mil traces but to use something thicker for the power connections. Do the routing on the underside, to keep the top surface of the design clean. The end result should look like Figure 8-13.

There is actually quite a lot of room on the left of the PCB, so a nice touch would be to put a table showing which Arduino pin is used for which lighting channel.

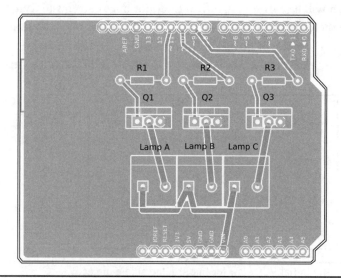

FIGURE 8-13   The final routed Arduino lighting shield.

Summary

In this chapter we have explored the use of Fritzing to support the use of Arduino and other prototyping boards.

In Chapter 9 you will learn how to modify and add parts to the Fritzing parts bin.

Custom Parts

Although Fritzing gives you access to a huge range of parts, inevitably you will discover a need for some part or other that is not available for Fritzing. You will also sometimes find that one of the parts included with Fritzing is almost what you need, but you need to alter it slightly. This is the case with the 555 timer IC used in the rear cycle lamp example, where the part name needed to be changed.

This chapter shows you how to create and modify parts for use in Fritzing. This process is complicated by the need to use a third-party graphics editor such as Inkscape to create and edit the graphics for the part.

In addition to editing and creating parts, there is always the option to search for parts that others have created and then published on the Internet as well as to use "generic" parts provided by Fritzing.

Generic IC Parts

You do not necessarily need to create a new part if Fritzing does not have the specific IC that you want to use in a project. Fritzing includes a generic IC part that you can use. You can specify the number of pins, the part name, and even the pin labels.

As an example, we could create the TLC555 timer chip that we need for the cycle lamp project as a generic part. Start by scrolling down the CORE parts bin until you get to the ICs section. Drag the first part in this section (just labeled IC) onto your Schematic view (Figure 9-1).

With the part selected, you can now customize this part. The default number of pins for this generic IC part is 8 pins, so we don't need to change that. However,

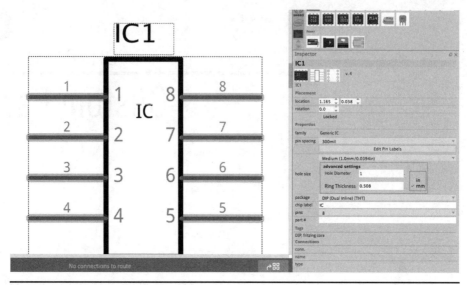

FIGURE 9-1   Adding a generic IC part.

if you had an IC with more pins, you could click on the Pins drop-down list in the Inspector. As you can see, there is no shortage of pin numbers to choose from. Similarly, you may also need to change the "package."

You will also need to change the field "part: #" to be the part number of your IC. In this case it is TLC555.

To change the pin names, click on the Edit Pin Labels button in the Inspector and the window shown in Figure 9-2 will appear.

Change these labels, using the datasheet of your IC as a reference, and then click on Save. The part on your schematic will then update, and you have a perfectly good part to use for your chip (Figure 9-3).

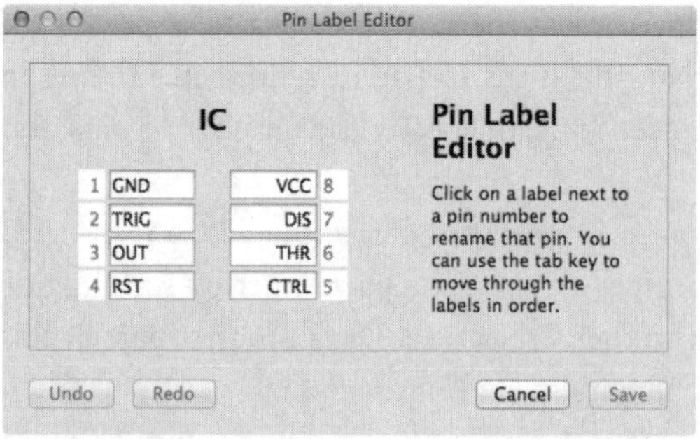

FIGURE 9-2   Changing the pin labels.

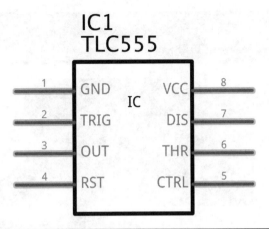

FIGURE 9-3 Making a TLC555 part from the generic IC part.

Finding Parts on the Internet

Another option that you can choose, before resorting to making your own part, is to see whether someone else has already made one and published it on the Internet. In fact Fritzing.org has promised a parts-sharing website. At the time of writing, this is not available, but it may well be by the time you are reading this.

Fritzing.org does maintain a list of parts that people have developed as part of their issue tracking system. If you visit https://code.google.com/p/fritzing/issues/detail?id=2753 and then search the page for the part you are interested in, you may well get lucky. The part will appear as a .fzpz file. Once it is downloaded, you can just double-click on the part file and it will be added to the Parts bin called MINE.

Alternatively, you can often get as good or better results simply by typing a query such as "HC-SR04 Fritzing part" into your favorite search engine. In this case, the HC-SR04 is an ultrasonic range-finder module. Searching for this turned up the Fritzing part in the Fritzing Projects area (Figure 9-4).

The downloads section over to the right of this page includes a download for the part. There are also lots of parts sitting in people's GitHub repositories that you can download and use. Be aware that sometimes the quality of these parts can be poor, so you may need to try out a few alternatives before you find one that's good.

FIGURE 9-4   Finding a part.

Modifying Parts

We have already fixed the problem of finding a part for the TLC555 by using a generic part. However, it's still not ideal, because the pins in the Schematic view are laid out in pin number order rather than being arranged by the pin's function (Figure 9-5).

The schematic symbol in Figure 9-5 labeled IC1 is the part you used when designing the cycle lamp example. The schematic symbol for IC1 has the pins organized by function, which is just what you want, except that the IC label is NE555 (and not editable) when we want it to be TLC555. The schematic symbol labeled IC2 in Figure 9-5 is the part you made earlier using the generic IC part, but this has the pins in package order.

To get the best of both worlds, you are going to edit the part labeled IC1 and change its part number. Unfortunately, this is not quite as simple as editing a field. The label NE555 is actually included in an image that you will need to edit.

Fritzing parts are based heavily on image files, and for best results, these image files are in the Scalable Vector Graphic (SVG) file format. Every Fritzing part has images for the schematic symbol, the breadboard part, and the part as it will

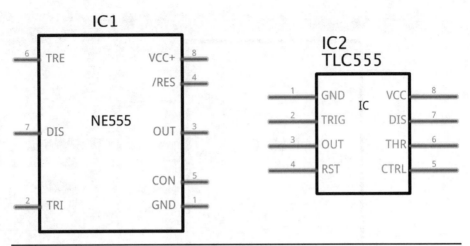

Figure 9-5   Schematic symbols by function and by package arrangement.

appear on the PCB view. It also needs an icon for the parts bin. These images are not just flat images, but make use of the concept of image objects to control things such as where connection points are. If you followed the description of creating PCB shapes using image files in Chapter 6, you will already have experienced editing such files using the open-source graphical editor called Inkscape. You are going to do the same thing here, to edit the part.

Add one of the NE555 parts to your Schematic view. Make sure you choose the one that looks like IC1 in Figure 9-5. Select the part and then choose the option Edit (New Parts Editor) from the Part menu. Switch to the Schematic tab of the parts editor (Figure 9-6).

If you hover over different areas of the image, in the parts editor, you will see that the pins and some other parts of the image will highlight. This reveals the different graphical objects in the image.

Over on the right-hand side of the window, you will find a Connectors section that defines the parts of the graphic that belong to a particular pin. You will notice that all the pin names are checked. This is so because those pins have parts of the graphic associated with them. We are going to replace the existing image, but only by editing the existing image, so that we will not disturb the names on the pin parts of the image that link to the names in the connector list.

Below the Connectors area of the parts editor window is an area labeled SVG. In here you will find the name of the SVG file being used by the Schematic view of the part. You are going to track down this file, make a copy of it, and then update the part to use the new image.

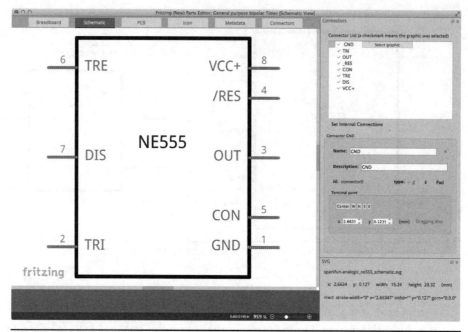

fritzing

FIGURE 9-6   Editing a part, Schematic view.

If you are using Windows or Linux, then using your operating system's feature to search for a file should find the file with as specific a name as "sparkfun -analog_ne555_schematic.svg" pretty easily. If you are a Mac user, then digging the file out is a little trickier as it will be embedded into your Fritzing app itself.

To find this file on a Mac, find the Fritzing app file in the Finder and right-click on it, selecting the option Show Package Contents. This will open a Finder window showing that an application file in OS X is actually a folder cunningly disguised as a file. Search within this folder for "sparkfun-analog_ne555_ schematic.svg." You should find it in parts/svg/core/schematic.

Whatever your operating system, you should now have the correct SVG file, so duplicate the file and name it "sparkfun-analog_**tlc555**_schematic.svg." Open the copy in Inkscape (Figure 9-7).

Double-click the text NE555 and modify the text to be TLC555. I also changed the font to be Arial, because when I didn't do this, I found that I got a warning from Fritzing about the font and then the text of the label shrank. Fritzing does recommend the use of certain fonts in their parts images, and you can read more about this here: http://fritzing.org/fritzings-graphic-standards. Save the file and quit Inkscape.

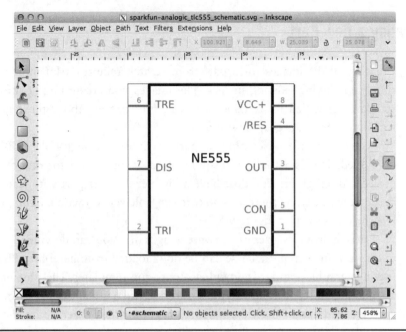

FIGURE 9-7 Editing a schematic part image in Inkscape.

Back in the parts editor on Fritzing, select the menu option Load Image for View from the File menu and select your newly edited SVG file. If you are using a Mac, you can select the file in the file chooser dialog by dragging the file in the Package Contents Finder window onto the file chooser dialog. Your parts editor should now show the modified part image for the schematic.

Click on the Metadata tab in the parts editor and change the "chip" property to "TLC555."

To save the new part, select the Save as New Part option from the File menu. Confirm the Prefix option, and now the part will appear in the MINE parts bin and the part on the Schematic view will also have updated to show the changed name.

If you want to share the part you have just modified, so that others can use it, then you need to export it by selecting the part on the Schematic view and then selecting the Export option from the File menu. For a chip like this, it's a good idea to just use the IC name as its file name, or in this case TLC555.fzpz. This file is then a complete self-contained part that can be added to the Fritzing parts list at https://code.google.com/p/fritzing/issues/detail?id=2753 or on your own blog or code repository.

Creating a New Part

Creating all the images for a part from scratch requires a lot more effort than simply modifying existing images. The images you create must conform to the standards specified by Fritzing. You can find out all about these here: http://fritzing.org/fritzings-graphic-standards/.

As an example of creating a part, a simple through-hole PCB buzzer is developed. This is actually in response to a genuine need for one of these in a project. Although Fritzing does have a big piezo buzzer part with leads included, at the time of writing there was no through-hole version with a 0.3 in pin spacing like the one shown in Figure 9-8.

Fritzing provides a set of template images that you can download and modify to create your own part. These can be downloaded from here: http://fritzing.org/fritzings-graphic-standards/download-fonts-and-templates. These assume that you are making an IC or module type of part and are quite a long way from what we need for the buzzer. To get the starting images for the buzzer, you will cannibalize a couple of different parts that are almost right.

Breadboard View

Figure 9-9a shows the breadboard view of an existing piezo buzzer on the left and a microphone part on the right, and Figure 9-9b shows the schematic symbols for the two parts.

Figure 9-8 A through-hole PCB buzzer.

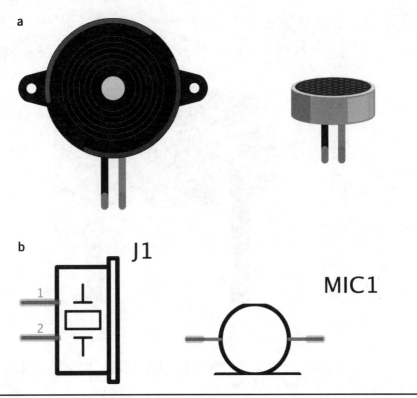

FIGURE 9-9 Piezo buzzer and microphone.

The package shape of the microphone is not a million miles from the buzzer that we want to model, and the schematic symbol for the piezo sounder is just right. So you can use the existing piezo sounder as the basis for your new part.

To get hold of the microphone breadboard image, drag the part from the parts bin onto the breadboard area. The part is called "Microphone" and is, for some reason, in the Output section of the CORE parts bin.

Looking at the SVG area of the parts editor, we can see that the image file for the Breadboard view is called just "microphone.svg." Using the same procedure as in the previous section, find the file (parts/svg/core/breadboard) and duplicate it, giving it the name "piezo_th.svg" ("th" for through-hole). Open the file in Inkscape and modify it by moving the red and black legs so that they are 0.3 in apart. To do this, you will need to ungroup some of the objects in the drawing. You could also change the dappled effect of the black top surface by selecting the top oval, ungrouping it, and then deleting all the tiny ovals that make up the dappled effect. Making the various parts of the cylinder darker will also make it closer in appearance to the actual component. Figure 9-10 shows the breadboard image being modified in Inkscape.

Figure 9-10   The modified breadboard buzzer image.

Because you are using a new breadboard image, after you have added it to the part using the Load Image for View menu option, you will need to select each of the pins on the right of the editor, click Select Graphic, and then click on one of the pins on the imported image. Figure 9-11 shows the graphic for pin1 being selected.

The schematic view for the original buzzer was fine, so there is nothing to change there; however, you need to do some work on the PCB view for the part. You can try out the part so far, just by saving a copy of the part and quitting the parts editor.

PCB View

The existing PCB image for the buzzer that we copied to make this part is not of much use here. First, the pads are the wrong spacing, and second, the part looks like a connector, because the original buzzer has wire on and is designed to be

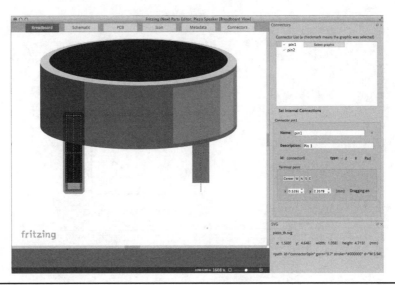

FIGURE 9-11   Selecting a graphic object for pin1 of the buzzer.

attached to a connector rather than be soldered directly to the PCB. So we will start from the template provided by Fritzing with the image for the PCB.

Make a copy of the template file PCBViewGraphic_Template.svg, call the copy pcb_piezo_th.svg, and open it in Inkscape to edit it.

To see the structure of this image, open the Inkscape XML Editor from the Edit menu. If you expand the folded code, you will see that among others there are sections with IDs of "silkscreen," "copper0," and "copper1." You are going to change what is in the "copper1" tag (bottom copper layer) and also change the silkscreen layer to be a circle rather than rectangular chip outline.

Changing the Copper

For this part, only two pads are needed. In fact, if we keep the top-left and bottom-left pads, these will be at the right spacing. So select each of the other tags in turn by selecting their line in the XML Editor view and then pressing the DELETE button (Figure 9-12).

When there just two pads left, center them horizontally as shown in Figure 9-13; then rename the two remaining pads to connector1pin and connector2pin.

Changing the Silkscreen

Now that the pads are okay, you need to work on the silkscreen. All graphic objects for the silkscreen should be white and have a stroke width of 0.75 pixel.

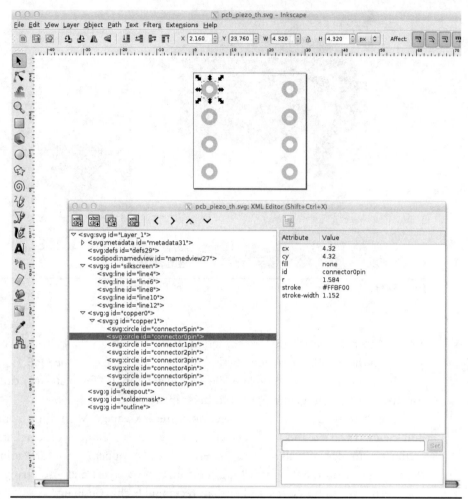

FIGURE 9-12   Rearranging the pads.

Since the document background is white by default, it's very difficult to see these lines. So set the document background to gray, using the option File | Document Properties | Background.

Now you can delete all the tags within the silkscreen tag (using the XML Editor), as you are going to replace them with a circle.

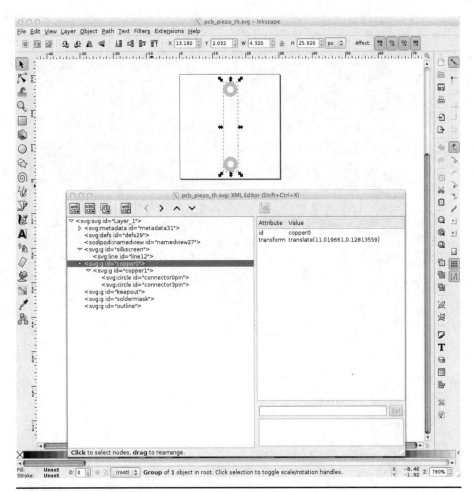

FIGURE 9-13   Centering the pads.

Add a circle using the Circle tool of Inkscape (Figure 9-14), adjust its size to fit the document bounds, and then make sure that its color is set to white (#FFFFFF) and its stroke width to 0.75px.

As you can see in Figure 9-14, the tag for the newly created circle is in the wrong place. It's at the bottom of the XML document but needs to be inside the silkscreen tag. Use the Raise Node and Indent Node buttons on the XML editor to move the circle (path tag) into the correct position, as shown in Figure 9-15.

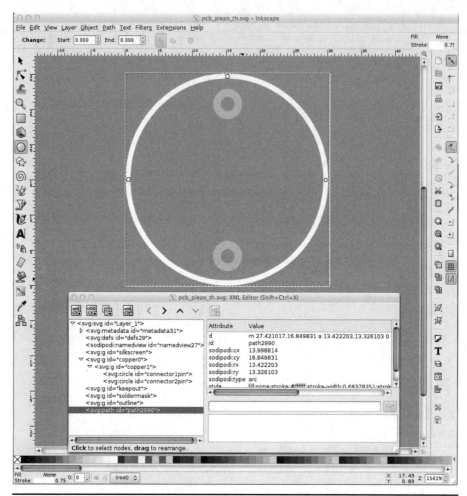

Figure 9-14   Adding a circle to the silkscreen.

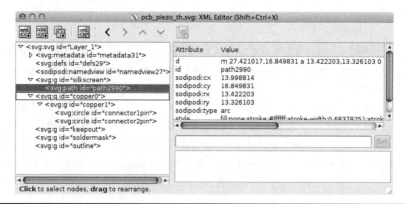

Figure 9-15   The silkscreen circle in the correct position.

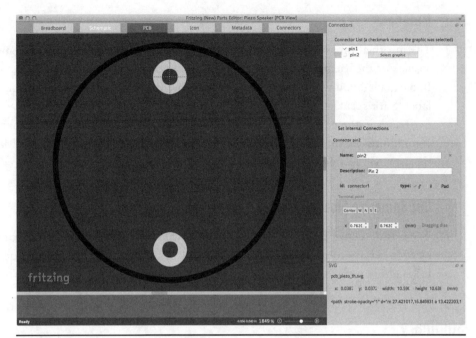

Figure 9-16 The modified PCB view image.

That completes the modifications to the image. You can save it and then switch back to Fritzing's part editor and select the option from the PCB view to use the new image, as you did with the Breadboard view. When you do so, the parts editor PCB view will look something like Figure 9-16.

You now need to connect the pins to the right parts of the PCB image, by selecting the pin in the right-hand side of the editor, and then click on the Select Graphic button. It does not really matter which pin you associate with which pad. When both pins are allocated a pad, they will both have a checkmark next to them.

Icon

The one tab that you have not yet visited in the parts editor is the Icon tab. This is just to set the icon that will appear in the parts bin. While you can create an image just for this, it makes sense to reuse the image used for the breadboard.

As with the other images, to change the image used for the icon, just select the menu option Load Image for View from the File menu and select the same image that you used for the breadboard.

Metadata

The information about the part (metadata) is not correct for this part any more, so click on the Metadata tab of the parts editor and make the necessary corrections. In particular, you will probably want to update the title, author, description, and label (S for speaker is better than J, which is usually used for connectors).

That's it, the part is now complete, and you can export it to share so that the community can make use of your work, as you did with the modified part from the previous section.

Get Someone Else to Do It

There is no denying that creating parts is a little tricky. For this reason, Fritzing offers a service to make parts for you. You can find out more about this on the Fritzing.org website.

Summary

Now that you can create your own parts to fill any gaps in Fritzing and the communities' parts library, there will be no stopping you.

In Chapter 10 you will find a number of example projects that you can use as guidance for your own projects.

Example Projects

In this chapter you will find a number of different projects that have been chosen to illustrate a particular area of electronics or that stretch the capabilities of Fritzing in some way. I will assume that by now you are familiar with Fritzing in general and are able to open and explore the example projects described here.

Henhouse Door

First, please don't try to build this project. As a design it has flaws and is an early prototype of the Arduino-based controller that I have to automatically let my hens in and out of their house. I have included it here as an example of how you can use Fritzing with stripboard.

Stripboard

Stripboard, sometimes called by the trade name of *Veroboard,* is a prototyping system that seems to be more popular in Europe than in the United States. Given the global nature of today's electronics industry, stripboard is also available in the United States.

Stripboard is a perforated PCB covered in holes at 0.1 in pitch. On the copper side of the board are strips of copper. This copper can be easily cut by using a special-purpose cutter (spot cutter) or by twisting a drill bit between your finders to break the track where needed.

Stripboard has the advantage of being more permanent and reliable than solderless breadboard at the cost of having to solder your design. For one-off designs they can be just as reliable as using a PCB. Other types of prototyping board are also available that have different patterns of strips on the underside of the board

The Stripboard Layout

Figure 10-1 shows the stripboard layout for this project. This uses the Breadboard tab, even though strictly speaking it's not using breadboard.

You can find this project in the downloads for the book. It has the name ch_10_Chicken_Door.fzz.

The copper strips would cause problems with the Arduino Pro Mini if they continued all the way across the board. If you drag the Arduino away from the breadboard a little, you can see that the copper strips have been cut (see Figure 10-2).

Notice how the air wires are showing where the Arduino needs to be connected on the stripboard. This is so because the Schematic view of the project (Figure 10-3) is influencing the Breadboard view, just as if you were using breadboard rather than stripboard.

Using Stripboard on the Breadboard View

Using stripboard on the Breadboard view is very similar to using breadboard. Start a new design and switch over to the Breadboard tab. Delete the breadboard

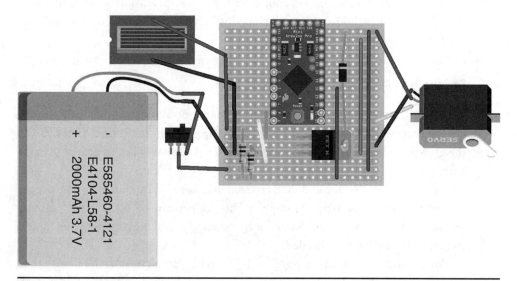

FIGURE 10-1 A stripboard project in the Breadboard view.

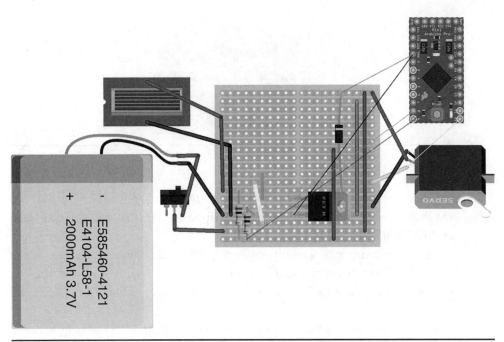

FIGURE 10-2   Stripboard layout with the Arduino moved to better see the cuts in the strips.

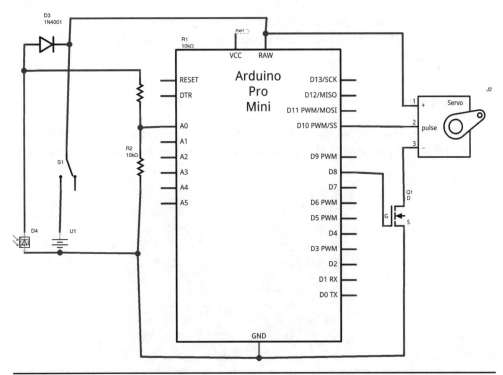

FIGURE 10-3   Schematic view of the stripboard project.

Figure 10-4   Stripboard part.

from the view, and then scroll down the CORE parts bin until you find the Breadboard section. Here you will find a stripboard part that you can then drag onto the Breadboard view (Figure 10-4).

The breadboard has a number of properties that you can change in the Inspector window when the stripboard is selected. In Figure 10-4, the size has been set to 10 columns by 7 rows, and the layout has been changed to horizontal strips.

Making cuts in the strips is simply a matter of clicking on the strip between two holes. To repair the cut, just click in the same place again. In real life, it is easiest to cut the strip over a hole, so it's usually best to make the cut on Fritzing on either side of the hole that is actually going to be cut.

Figure 10-5 shows cuts made in the center of the strip on the top row of the stripboard.

Perfboard

You may have noticed another option next to stripboard in the parts bin called *Perfboard*. Perfboard is like stripboard but without the strips. Some types have

Figure 10-5   Cutting the strips in Fritzing.

solder pads under each hole that are not connected, but serve to anchor the components. Other types of Perfboard are just a board with holes in it. Either way, the connections between components are often made by bending the component leads and soldering them on the underside of the board.

Example: Breakout Board

Many new designs use surface-mount ICs that do not have through-hole counterparts, which makes prototyping with breadboard tricky. A common technique to allow SMDs to be used on breadboard is to use a *breakout board*. These boards "break out" the surface-mount pads into standard 0.1 in header pins so that they can fit onto your breadboard.

This example is a breakout board for the L9110 low-cost H-bridge for motor control. The device is only available in SO08 (Small Outline 8 pin), and as well as holding the chip, the board will add a decoupling capacitor.

Schematic Design

Start by adding a generic IC and then change the pin labels as described in Chapter 9 so that they look like Figure 10-6. To add the header pins, find the Connection section of the CORE parts bin and then select Generic Female Header 2 pins and drag it onto the schematic.

Actually, you need a male header with 6 pins, so select the header that you just added and in the Pins drop-down menu select 6, and then in the Form drop-down menu select male.

Next add a 100nF 16V capacitor in a 0805 SMD package. Finally, connect everything as shown in Figure 10-6 and then add labels to the connector as six separate Text parts.

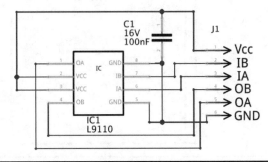

Figure 10-6   Schematic for a L9110 breakout board.

PCB Design

The whole point of this project is to allow an SMD to be used on breadboard, so there will be no actual Breadboard view for this project. Turn to the PCB tab and drag all the components onto the PCB area. Note that sometimes when you add a part, it may be way off to one side of the PCB area and you may think the part has not actually been added. A good tip for finding such parts is to zoom right out until you find it.

Rearrange the parts on the PCB and then resize it to just contain the parts with a little room around it (Figure 10-7).

Looking at Figure 10-7, you can see that the air wires for the two outputs OA and OB cross each other. This would make it difficult to route them without using a via, and there is no particular reason why they have to be in that order on the schematic. So at this point, jump back to the schematic and swap the connections over on the pin header J1, as shown in Figure 10-8. Remember to alter the labels next to the pin header.

Now, when you return to the PCB view, the output air wires no longer cross.

Start routing on the top layer, dropping down to 16 mil trace widths for the immediate connections to the IC pins. The power connections to C1 can be 24 mils. Add in one via for V_{CC}. The segment of trace from the via to IC1 is 16 mils, and the bottom side trace to J1 is 24 mils.

Add in text to appear on the silkscreen layer that identifies the pins of the header and also names the board. When you add the text for the pins, make one label and then duplicate it to keep a consistent text size. The text is resized by dragging the corners of its enclosing rectangle. The end result is shown in Figure 10-9.

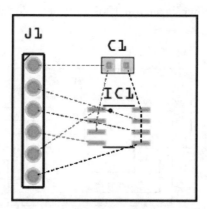

Figure 10-7   Breakout board with parts positioned.

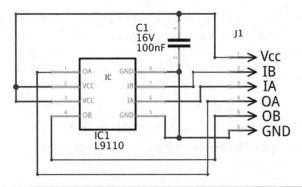

FIGURE 10-8   Swapping the output pins.

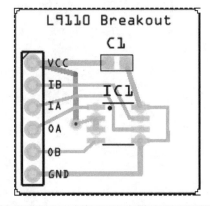

FIGURE 10-9   Final board layout.

You may wish to further improve the design with holes for mounting the board, copper fills, and curved edges to the PCB.

Examples on Fritzing.org

As well as the examples here, there are Fritzing designs that come bundled with the tool. These can be accessed directly from the Open Example option on the File menu. Many of these relate to breadboarding with Arduino, rather than taking the extra step to PCB design, but some are worth exploring:

- Matrix—an LED matrix Arduino shield
- Accelerometer—Arduino shield using an accelerometer
- Voltage regulator with switch

There are also links to other people's projects as well as projects made by the Fritzing team at http://fritzing.org/projects.

Summary

In this chapter you have seen how Fritzing can be used for a variety of different types of project. In Chapter 11 we will explore testing and ways to make your product reliable.

Testing

The final two chapters of the book are devoted to the more peripheral aspects of developing indie hardware. In this chapter, we look at testing, an area that gets much attention in professional development, but is often neglected in the "maker" domain.

If you are just making something for your own personal use, then testing does not need to be any more than trying it and seeing if it works. However, if you plan to make a product to sell, then you need to be sure that the design is correct before you have a whole load made.

What is more, if your design is being made by someone else, who is also going to sell the boards on your behalf, then the seller needs to be able to test each of your products before they leave the shop.

Testing electronics thoroughly is a discipline in its own right, and so we can really touch on only the basics here.

Reliability

Before we look at how to test your design, it is worth ensuring that we have done everything possible at the design stage to ensure the product can be manufactured reliably and will not fail soon after someone starts using it.

If you make yourself one of a thing and it works, you might assume that if you make 1000, then they will all work too. This is not the case. You need to be more careful than that. There are all sorts of reasons why some or even most of them might not work. For example, the manufacturer may get the parts from a different supplier that has slightly different behavior, which might be enough to stop the project working. Or, you might have forgotten to update the BOM after substituting a component late in the design.

Guidelines

Here are some easy ways for getting a reliable design:

- Build one from the BOM that you are giving to your supplier, using the design you are giving the supplier. There should be no shortcuts or exceptions and no building from memory.
- Use the reference designs in datasheets and read the datasheet from end to end, especially the application notes. Pay particular attention to any section on selecting component values and types.
- Don't stretch component tolerances. If your design is going to operate at 9V, then use 16V capacitors not 10V ones. Similarly, if a resistor is going to be burning ¼W of heat, then use a ½W resistor.

Capacitors

Most components will last longer than people are reasonably likely to want to use your product, but the exception to this is some types of capacitor.

The most commonly used capacitors are multilayer ceramic (MLC) and electrolytic. MLC capacitors are very reliable: they will operate for many thousands of hours unless you abuse them by exceeding their voltage or temperature ratings. Electrolytic components, however, are notoriously unreliable. Many are only good for 5000 hours of operation, being pushed hard, but within their specification. That's actually less than one year. Their capacitance will gradually decrease and then they will fail.

Any repairer of old audio equipment will tell you that the first thing to do is to replace all the electrolytic capacitors. It used to be that any MLC capacitors were only available up to capacitances of less than 1μF or so without being prohibitively expensive. Improvements in the technology mean that MLC capacitors are now available at the hundreds of microfarads range at little greater cost than their electrolytic equivalents. This effectively leaves electrolytics to the domain of high-capacity (above 100μF) applications such as smoothing power supplies.

Tantalum capacitors are a different technology from electrolytic and are commonly used as more expensive replacements for electrolytics at 1 to 100μF. As well as being more expensive than electrolytic, they have the unfortunate habit of failing to a short circuit, with fiery results. I would consider these best avoided in favor of high-capacity MLC capacitors in new designs.

Transistors

If you are working with digital designs, then you are likely to be using transistors as switches.

Bipolar Transistors

Bipolar transistors will not generally cause you any trouble. As with everything, it is essential to ensure that the tolerances of power, current, voltage, or temperature are not exceeded. One area that you should also consider, when using a bipolar transistor as a switch, is that this kind of transistor amplifies current by a fixed amount, called its *gain*. Figure 11-1 shows how you might use a transistor to control a motor using an Arduino output pin.

The maximum current flowing through the motor, then through the collector and emitter of the transistor, is the current flowing through R1, and the base to emitter of the transistor multiplied by the gain. So if the transistor had a gain of 100 and if 1mA flows through R1, then the maximum current flowing through the motor is 100mA.

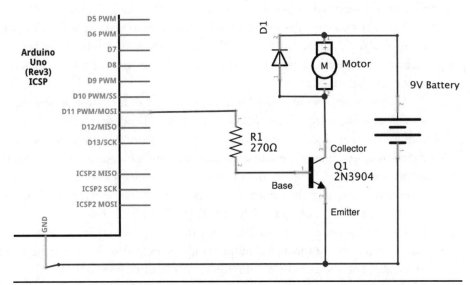

FIGURE 11-1   Controlling a DC motor with a transistor.

Let's now look at a scenario where not paying sufficient attention to this could cause us problems.

There is always a voltage drop of about 0.5V across the base and emitter of a transistor, so the current when D11 is high (5V) will be $I = V/R = 4.5/270 = 16$mA. For a transistor with gain of 100, the maximum current through the motor would be (given a high enough battery voltage) a very short-lived 1.6A. I say short-lived because the transistor can only cope with about 200mA. So, in short, this design would be fine even if the transistor gain were only, say, 20.

Transistor gain can vary wildly between batches. If you look at the datasheet for a 2N3904, the gain (called hfe on the datasheet; see www.fairchildsemi.com/datasheets/2N/2N3904.pdf) is specified as somewhere in the range of 60 to 300. So, if a higher value of R1 were used and we happened to have an individual transistor with a gain of 300, all might be well—until the manufacturer of our boards used a batch of transistors with a gain of 60 (still within specification) and then the design would not work.

In short, don't rely on luck; do the math.

MOSFETs

N-channel MOSFETs do not suffer from the need to supply enough controlling current, but they do have some design parameter ranges that you need to be aware of. First, there is the gate threshold voltage (VGS). This is the voltage that the gate must be above for the MOSFET to turn on. If you look at the datasheet for an FQP33N10 (www.fairchildsemi.com/datasheets/FQ/FQP33N10.pdf), you can see that VGS is specified as having a minimum of 2V and a maximum of 4V.

Let's say you were designing a project that uses a 3.3V logic level to switch a transistor. Well, judging by the datasheet, it's almost 50:50 that 3.3V will not be sufficient to switch the transistor. Just because your MOSFET works, there is no guarantee that another will.

Second, when you use a MOSFET, note the use of a resistor between a digital output and the gate of the MOSFET. Often DIY designs omit this, instead connecting the digital output directly to the gate. The resistor is generally considered good practice, because the gate and source connections of the MOSFET act as a capacitor, so when the digital output switches between low and high, current will rush into the capacitor as it charges up. The capacitance is not large, perhaps a maximum of 2 to 3nF for the FQP33N10. But without any resistor and a very small amount of resistance from the PCB trace (if it is very short), the current capability of the digital output could be exceeded when switching at high frequency. For the cost of a cent for a 1kΩ resistor, this potential problem is eliminated.

Design for Testability

When you design your project, it's a good idea to have this question lurking somewhere at the back of your mind: How am I going to test this? In many cases, you will be able to make a test harness (see the section "Automated Testing" later in this chapter), but this can only work if you can get electrical connections everywhere that you need to.

Figure 11-2 shows the schematic of Figure 11-1 but with a test point part (labeled TP1) connected to R1 and the Arduino's digital output. This test point will appear on the PCB view as a pad. You can then touch a multimeter probe or probe from an automated test harness to this point to measure the voltage. You can find the test point part if you just search for "Test Point" in the parts bin.

Functional Testing

The most obvious type of testing is functional testing. This answers the simple question, Does it work? In actual fact what you want to know is, Does everything work? This means pressing every button, adjusting every control that can be adjusted, and observing the effect on the system. It can really help to write yourself a short test script; this can be a short to-do list that just ensures that you don't miss anything.

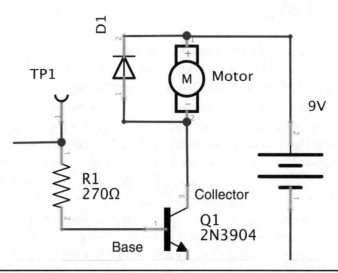

FIGURE 11-2   Test points.

Electrical Testing

The current consumption of a project is a really good indicator of the health of your project, so use a multimeter or power supply with a built-in ammeter. It's very similar to measuring a patient's blood pressure. If you are hemorrhaging current, you need to find out why.

Before switching on, decide what the current should be. Then turn it on and see if you are right. If you do it the other way around, it's very easy to convince yourself that things are okay when they aren't.

If your design is drawing more current than you would expect, try to work out why this is. Measuring voltages all over the project, starting with the supply rails, is always a good idea. Also check to see if any of the components are getting hotter than they should. A thermometer will help you be objective about this.

Just because a project works for 10 minutes while you try it out does not mean that it will work indefinitely. Before declaring a project to be good, you need to leave it running for as long as your patience will allow. Once it has been running for an hour or so, check all the components to see if anything is getting hot. Some components will get hot, especially transistors switching large currents, resistors, and voltage regulator ICs. If you have any doubts, check these temperatures with a thermometer and make sure that they do not exceed the maximum temperature for the component.

Automated Testing

If you are having someone else do your manufacturing, one of the things that you may well be asked for is a test harness so that they can check every individual item as it is produced. This can be a very time-consuming thing to do by hand, and the more labor-intensive it is, the more it will add to the cost of production.

Some products are easily tested by connecting a power supply, turning it on, checking that it does what it's supposed to, disconnecting it, and sending it off to packaging. This might only take 10 seconds per item. Other projects can have a lot of possible things to test. As an example, Figure 11-3 shows a test framework that I developed for a Raspberry Pi expansion board. Initially, this product was being tested by plugging it into a Raspberry Pi and running a test program, that took considerable time, especially as the Raspberry Pi needed to reboot each time.

In the end, I developed an Arduino-based test harness that used the Arduino pins connected to all the inputs and outputs of the product. This reduced the testing to a series of LED flashings and blinkings to indicate success or failure.

FIGURE 11-3 An Arduino-based test harness.

Since there would only ever be one of these test harnesses, as you can see, it was built using stripboard and an Arduino Protoshield rather than have its own PCB designed for it.

Summary

In this chapter, you have only touched on the basics of electronics testing. Always remember just how difficult a product recall would be, and do all you can to ensure your design is watertight.

CHAPTER 12

Funding and Distributing

C rowd-sourced funding offers an opportunity for the inventor to both fund and publicize an invention. To us technical types, the business aspects of such projects are often considered trivial, but actually this is not an area that can be ignored.

Secrecy Versus Openness

This chapter does not talk about patents and other ways of protecting intellectual property. It probably should, and I encourage you to look elsewhere for information on this topic. This omission has two reasons. First, I have no experience in this area, and second, I think that an open-source model for hardware design has a lot to offer.

The traditional way for an inventor to proceed with his or her invention would be in the utmost secrecy, telling no one about the invention without a lot of fuss and nondisclosure agreements (NDAs) and sometimes a fair degree of paranoia. At the same time a patent application would be drawn up at great expense with the help of a patent lawyer to capture the essence of your "invention" and protect it from being copied by others.

Most inventors believe that their idea is valuable and unique. If they didn't, then they would not pour time and effort into it. However, discussing an idea with others is incredibly valuable and helps to put things into perspective. True, you must choose your advisers carefully. They need to be people who know as much as or more about the project than you. You probably don't need to worry about them stealing your idea, because however good it is, they will probably believe that they have better ideas of their own that they would like to exploit—if only they could find the time.

There are, I am sure, exceptions to this. The truly radical idea that somehow everyone missed is worthy of great caution. Perhaps James Dyson's cyclone system for vacuum cleaners is one such revolutionary invention. But most inventions are incremental changes to something that already exists. I find it not altogether fair to try to protect an idea that lots of people have already had.

Open-Source Hardware

In recent years, open-source software has become more and more popular. Linux is of course the most prominent of open-source projects. The availability of source code, warts and all with no secrets, is immensely appealing. If you have a problem with your open-source software, then you can fix it yourself if you want. Obviously, that requires a fair investment in time and skills, but at least it's an option.

Open-source hardware (OSH) follows a similar model for electronic design. Producing an OSH product involves releasing the design files for the project, schematic, board design, source code to any firmware, etc. This puts the design under scrutiny and allows the community to suggest improvements to your design and, as a community, to promote your product and self-support. Plus there is the kudos associated with appearing to do the right thing and make the project open-source.

The publisher Tim O'Reilly has a great catch phrase: "Create more value than you capture." For me, this sums up the OSH movement. Yes, there is a risk that your design will be copied. Clever parts of your design are likely to be copied. Indeed, you can take clever ideas from other OSH products and use them in your OSH product. However, you have to consider that even if your design is copied, most new businesses only ever achieve a tiny share of what is a global market. But in the meantime, your product is being popularized.

Kickstarter

At this point, I should point out that there are many crowd-funding platforms out there, of which Kickstarter is just one. They all have slightly different takes on crowd funding.

Kickstarter began as a crowd-funding tool to support artistic and creative endeavors. The basic idea is that anyone, individual or company, could make a pitch for a project idea that needed some money to get it off the ground. Rewards would be offered to people backing the project, but giving equity as a reward was not an option. A valid type of reward might be a front-row seat at a first performance of a play (if that was the project), or a more appropriate example might be an actual electronic product from the first batch of products.

A target sum is set by you. Then for a window of typically one month, people would buy your rewards in order to back your project. If you exceed the target amount, then some money lands in your bank account and you have the responsibility of fulfilling all the promises you made as rewards in a timely manner. If the target is not reached, then you don't get any of the money.

As a way of funding the development and first-run costs of a new electronics project, this has a number of advantages:

- You don't have to use your own money or go to a bank for a loan.
- The Kickstarter campaign will generate lots of interest in your product (if it's any good).
- Some of the rewards that you offer will effectively be preorders for your first production run. In other words, it removes a lot of the risk from the whole process of getting a new product to market.

You will find plenty of information on www.kickstarter.com about creating a successful project, but here are some key things to remember:

- Do the math! Make sure that your target will cover the costs of giving out the rewards.
- Set realistic time scales. It is always better to underpromise and overdeliver. People have been kind enough to use their own money to help you out. Don't let them down because you underestimated how long things would take.

Seeed Studio and Sparkfun

Seeed Studio (yes, there are three e's) describes themselves as an OSH facilitator. They offer a wide range of services: bare PCB manufacturing and PCB prototyping through to full production and test. They will also sell your design retail or allow other retailers to buy your product from a wholesale portal at a different price from the retail price. This means that a lot of the tedious business side of electronic invention is managed by someone else who just sends you some money every now and then.

Sparkfun, and I am sure other electronic hardware retailers, will also enter into similar arrangements, often on a royalty basis, where after the initial design all further manufacturing, distribution, and retail are handled by them.

Summary

You can have the best electronic invention of all time, but it will come to naught without the ability to address the business side of things. In this chapter, I have looked at this issue from the standpoint of an electronics enthusiast rather than someone hoping to create a business. If you are more interested in making inventions than making money, then you may find it beneficial to team up with someone who is more interested in the business side of things.

Resources

For latest information and resources, there is nothing better than spending some time with your favorite search engine. However, here are a few useful resources that I have collated. I have not tried many of these organizations, so do not consider these to be recommendations or endorsements on my part. As always, search out other people's experiences on the Internet.

Bare PCB Manufacturers

Here are some other (apart from Fritzing.org) manufacturers of prototype PCBs.

Company	Website	Notes
Seeed Studio	http:www.seeedstudio.com/service/index.php?r=pcb	Fusion PCB service. From $10 for 10 boards.
OSH Park	https://oshpark.com/	Made in the United States. From $5 per square inch.
ITEAD Studio	http://imall.iteadstudio.com/open-pcb.html	Open PCB service. Low cost.
Elektor	http:www.elektorpcbservice.com/	European service.

Prototyping Services

When it comes to prototyping services, both Seeed Studio and ITEAD Studio offer cusomized services.

Company	Website	Notes
Seeed Studio	http://www.seeedstudio.com/service/index.php?r=pcb	From $20 per assembled board.
ITEAD Studio	http://imall.iteadstudio.com/open-pcb.html	Open PCB service. Low cost.

Adafruit also maintains a useful list of PCB manufacturers at www.ladyada .net/library/pcb/manufacturers.html.

3D Design and Printing Services

This is probably the most rapidly growing area, and it is definitely worth checking out the latest situation. Here are some companies offering such services.

Company	Website	Notes
Sculpteo	http://www.sculpteo.com/	Printing in 48 different materials.
Shapeways	http://shapeways.com	Ships from United States and Europe.

Component Suppliers

Company	Website	Notes
Digikey	http://www.digikey.com/	
MCM Electronics	http://www.mcmelectronics.com/	
CPC	http://cpc.farnell.com	Based in United Kingdom.
Farnell	http://www.farnell.com	International.
Maplin	http://www.maplin.co.uk/	Brick and mortar shops based in United Kingdom.

Common Resistor and Capacitor Values

Certain values of resistors and capacitors crop up time after time. These are components that I like to keep a good stock of. The most common resistor values that you are likely to encounter are 10, 100, 220, 270, and 470Ω; 1, 10, and 100kΩ; and 1MΩ.

You can generally get away with an even more restricted range of capacitor values: 10 and 100nF, and 10 and 100μF.

Index

References to figures are in italics.